Room-temperature Sodium-Sulfur Batteries

This book provides an effective review and critical analysis of the recently demonstrated room-temperature sodium-sulfur batteries. Divided into three sections, it highlights the status of the technologies and strategies developed for the sodium metal anode, insight into the development of sulfur cathode, and electrolyte engineering. It reviews past, present, and future perspectives for each cell component including characterization tools unveiling the fundamental understanding of the room-temperature sodium-sulfur batteries.

FEATURES:

- Highlights scientific challenges in developing room-temperature sodium-sulfur batteries
- Covers pertinent anode, cathode, and electrolyte engineering
- Provides scientific and technical interpretation for each of the cell components
- Discusses how Na-S batteries relate to the more extensively researched Li-S batteries
- Explores importance of the SEI and CEI in developing stable sodium-sulfur batteries

This book is aimed at graduate students and researchers in energy science, materials science, and electrochemistry.

Room-temperature Sodium-Sulfur Batteries
Anode, Cathode, and Electrolyte Design

Edited by
Vipin Kumar

CRC Press is an imprint of the
Taylor & Francis Group, an **informa** business

Designed cover image: © Vipin Kumar

First edition published 2024
by CRC Press
2385 NW Executive Center Drive, Suite 320, Boca Raton FL 33431

and by CRC Press
4 Park Square, Milton Park, Abingdon, Oxon, OX14 4RN

CRC Press is an imprint of Taylor & Francis Group, LLC

© 2024 selection and editorial matter, Vipin Kumar; individual chapters, the contributors

Reasonable efforts have been made to publish reliable data and information, but the author and publisher cannot assume responsibility for the validity of all materials or the consequences of their use. The authors and publishers have attempted to trace the copyright holders of all material reproduced in this publication and apologize to copyright holders if permission to publish in this form has not been obtained. If any copyright material has not been acknowledged please write and let us know so we may rectify in any future reprint.

Except as permitted under U.S. Copyright Law, no part of this book may be reprinted, reproduced, transmitted, or utilized in any form by any electronic, mechanical, or other means, now known or hereafter invented, including photocopying, microfilming, and recording, or in any information storage or retrieval system, without written permission from the publishers.

For permission to photocopy or use material electronically from this work, access www.copyright.com or contact the Copyright Clearance Center, Inc. (CCC), 222 Rosewood Drive, Danvers, MA 01923, 978- 750-8400. For works that are not available on CCC please contact mpkbookspermissions@tandf.co.uk

Trademark notice: Product or corporate names may be trademarks or registered trademarks and are used only for identification and explanation without intent to infringe.

ISBN: 9781032456089 (hbk)
ISBN: 9781032482422 (pbk)
ISBN: 9781003388067 (ebk)

DOI: 10.1201/9781003388067

Typeset in Times
by codeMantra

Contents

Preface...ix
Acknowledgments..xi
Editor's biography..xiii
List of contributors..xv

Chapter 1 Introduction: sodium-sulfur battery technology........................1

  S.K. Vineeth and Vipin Kumar
  1.1 Introduction: sodium-sulfur battery technology.....................1
  1.2 Brief history of Na-S battery..2
  1.3 Sodium metal batteries and the operation of HT Na-S batteries...3
  1.4 The transition from high-temperature to RT Na-S batteries.....5
  1.5 Development of the RT Na-S batteries..................................9
  1.6 Conclusion and prospects...12
  References..12

Chapter 2 Sodium metal anode: past, present, and future of sodium metal anode...17

  Chhail Bihari Soni and Vipin Kumar
  2.1 Introduction to the sodium metal anode...............................17
  2.2 Challenges in developing a stable sodium metal anode.........19
    2.2.1 Unstable solid electrolyte interphase........................19
    2.2.2 Dendrite growth..20
    2.2.3 Instabilities due to localized physical changes..........21
    2.2.4 Instability due to localized chemical changes..........22
  2.3 Strategies to overcome the challenges.................................22
    2.3.1 Surface polishing..23
    2.3.2 Electrolyte engineering...23
    2.3.3 Interface engineering – designing intrinsic and extrinsic interphases...25
    2.3.4 Nanostructured host and current collector................28
  2.4 Future prospects..28
  References..35

Chapter 3 Sulfur cathode: progress in the development of sulfur cathode.........39

  Sungjemmenla, S.K. Vineeth, and Vipin Kumar
  3.1 Introduction to sulfur cathode..39
    3.1.1 Mechanistic principle of the sulfur conversion reaction...39

	3.2	Challenges in developing the stable sulfur cathodes 41	
		3.2.1	Insulating nature of sulfur and its discharge species 41
		3.2.2	Polysulfide shuttling 42
		3.2.3	Volume expansion 42
		3.2.4	Loss of sulfur species 42
		3.2.5	Self-discharge 42
		3.2.6	Low Coulombic efficiency 43
	3.3	Progress in developing sulfur cathodes 43	
		3.3.1	Physically adsorbed S-host 43
		3.3.2	Chemisorbed S-host 46
		3.3.3	Covalently bonded S-host 52
		3.3.4	Hybrid S-host 54
	3.4	Importance of electrolyte/sulfur (E/S) ratio in developing a stable sulfur cathode 55	
	3.5	Strategies to develop high-loading sulfur cathodes 56	
		3.5.1	Synthesis techniques 57
		3.5.2	Coating techniques 58
	3.6	Future prospects 64	
		3.6.1	Advanced characterization techniques 65
		3.6.2	Computational models 66
	References 68		

Chapter 4 Electrolytes for room-temperature sodium-sulfur batteries: a holistic approach to understand solvation 79

SK Vineeth, Sungjemmenla, Yusuke Yamauchi, and Vipin Kumar

	4.1	Basic properties of the electrolytes for alkali metal batteries 79	
		4.1.1	Evolution of the room-temperature sodium-sulfur battery – transition from a solid-electrolyte to the liquid electrolyte 81
	4.2	The solid-electrolyte interphase (SEI) and the importance of sodium-ion solvation 85	
	4.3	Liquid electrolyte to quasi-solid-state electrolyte to solid electrolyte for sodium-sulfur batteries 86	
		4.3.1	Advanced liquid electrolytes for RT Na-S batteries 88
		4.3.2	Quasi-solid-state electrolytes 92
		4.3.3	Solid-state electrolytes 93
	4.4	Nature of electrolyte and its role in developing a stable SEI and CEI 102	
	4.5	Future prospects 104	
	References 105		

Contents

Chapter 5 Analytical techniques to probe room-temperature sodium-sulfur batteries 115

Chhail Bihari Soni, Sungjemmenla, and Vipin Kumar
5.1 Introduction 115
5.2 Overview of the routine techniques to probe sodium metal anode 116
5.3 Overview of the routine techniques to probe sulfur cathode 117
    5.3.1 X-ray diffraction 118
    5.3.2 UV-vis spectroscopy 119
    5.3.3 Raman spectroscopy 119
    5.3.4 X-ray absorption near edge structure 120
    5.3.5 X-ray photoelectron spectroscopy 121
5.4 In situ/operando techniques for sodium-sulfur batteries 122
    5.4.1 In situ OM 122
    5.4.2 In situ atomic force microscopy (AFM) 123
    5.4.3 In situ electron-microscopy techniques 123
5.5 Future prospects 123
References 124

Chapter 6 Sodium-sulfur batteries: similarities and differences with lithium-sulfur battery 127

C. Sanjaykumar, Rahul Singh, Rajendra Singh, and Vipin Kumar
6.1 Na-S and Li-S chemistries are fundamentally different – a mechanistic overview 127
6.2 Polysulfide species dissolution 131
6.3 Electrolyte for Li-S and RT Na-S batteries 131
    6.3.1 Solid electrolyte interphase 132
6.4 Future prospect 134
References 134

Chapter 7 Other sodium metal-based rechargeable battery technologies: A brief introduction to sodium dual ion batteries 137

Chhail Bihari Soni, SK Vineeth, and Vipin Kumar
7.1 Bird-eye view of the sodium metal-based batteries 137
7.2 Overview of the sodium-air battery 139
7.3 Overview of sodium dual-ion batteries 141
    7.3.1 Overview and history of dual ion batteries 142
    7.3.2 Developments of sodium dual ion batteries 143
7.4 Future prospects 146
References 146

Chapter 8 Conclusion and prospects 151
Vipin Kumar
    8.1   Conclusion and prospects 151

Index 153

Preface

In 1966, a team from Ford Motor's research division disclosed the details of a new and innovative battery concept they had developed. The battery consists of sodium metal anode and sulfur cathode in the liquid state and a solid electrolyte membrane. The solid electrolyte membrane, i.e., a ceramic electrolyte comprising over 90% of aluminum oxide, is permeable to sodium ions while blocking the movement of sulfur and its discharge products. This new finding brought together materials scientists, electrochemists, physicists, mathematicians, and engineers from Ford Motor to further develop the battery technology for mobility applications. However, due to its high operating temperature, sodium-sulfur battery chemistry was never deployed for electric mobility applications. Since then, there has been a dire need to develop high-energy, safe, and long-lasting sodium-sulfur batteries. Li-ion battery chemistries have played a vital role in the development of electric mobility and other energy insatiable applications ever since their launch in 1991. Despite their low specific energy density and limited drive range compared to gasoline (i.e., 10–12 kWh/kg), Li-ion batteries have acquired ample space in mobility. To achieve a more extended drive range, one must explore new battery chemistries different from the existing Li-based ones.

One such battery chemistry is room-temperature sodium-sulfur battery technology; the operating principle and operation mechanism are similar to that of the high-temperature sodium-sulfur battery, which has been known for almost six decades. In principle, a room-temperature sodium-sulfur battery can satisfy all the basic requirements of the intelligent vehicle battery system since it offers a high specific energy density and is lightweight compared to lithium-iron-phosphate battery chemistry. The unique attributes of room-temperature sodium-sulfur batteries in capacity, energy density, and price are apparent compared to Li-based battery chemistries. Room-temperature sodium-sulfur battery chemistry deserves a careful investigation to provide evidence for its commercial viability.

This first book on room-temperature sodium-sulfur batteries provides the reader (researchers, industries, academicians, etc.) an excellent review and critical analysis of the recently demonstrated room-temperature sodium-sulfur batteries. This book is divided into three sections. The first section highlights the status of the technologies and strategies developed for the sodium metal anode. The second section provides a detailed insight into the development of sulfur cathode. The last section is devoted to unveiling electrolyte engineering. Every section starts with a brief history of the topic and the related chemistry and underlying challenges. This book provides past, present, and future perspectives for each cell component, i.e., sodium metal anode, sulfur-based cathodes, electrolytes, and separators. In addition, the characterization tools unveiling the fundamental understanding of the room-temperature sodium-sulfur batteries have been discussed in the end. We recommend this book to energy engineers, scientists, and industries.

This book on room-temperature sodium-sulfur batteries provides readers with an excellent collection of progress and development on the battery components, i.e., sodium metal anode, sulfur cathode, and electrolyte. The authors of this are deeply involved in research for the progress of room-temperature sodium-sulfur batteries and concentrate on building a fundamental understanding of the electrochemical processes. The crucial challenges while working with the cells are detailed with possible solutions. As the editor, I would like to thank and appreciate all the contributing authors for their outstanding efforts in making this book a reality.

Acknowledgments

Besides the authors of this book, many people provided assistance that helped immensely in bringing this edition of the book to realization.

I am most grateful to the following graduate students of mine for the time and effort they dedicated to assist in the preparation of the figures for chapters and cover image:
- Mr. Vineeth SK
- Mr. Sanjaykumar C
- Mr. Rahul Singh
- Mr. Lovlesh Roy

Their assistance helped me enhance the content and make it more focused as a comprehensive resource and a valuable book on room-temperature sodium-sulfur batteries.

I acknowledge the support provided by my current academic institution.

Also, I sincerely appreciate the encouragement of the publisher and their recognition of the increasing importance of room-temperature sodium-sulfur batteries.

Editor's biography

Dr. Vipin Kumar is an Assistant Professor in the Department of Energy Science and Engineering (DESE) at the Indian Institute of Technology Delhi (IIT Delhi), India. He is the coordinator of a virtual interdisciplinary school, i.e., the School of Interdisciplinary Research (SIRe) at IIT Delhi.

He received his doctoral and master's degrees from Nanyang Technological University Singapore (NTU Singapore) and IIT Delhi in 2016 and 2011. He worked with the Institute of Materials Research and Engineering (IMRE), Agency for Science, Technology, and Research (A*STAR), Singapore, to develop high-energy metal-sulfur batteries. His research focuses on electrochemical energy storage devices, such as room-temperature metal-sulfur batteries (e.g., Na-S, Li-S, Al-S, and Mg-S).

Most of his studies were devoted to energy-related projects. He has received several prestigious awards, including the "Inspire Faculty Award" offered by the Department of Science and Technology, India, in 2017. Since May 2020, he has been working as an Assistant Professor at the Indian Institute of Technology Delhi (IIT Delhi), India, and running his battery research group (Advanced Batteries Research Laboratory) in the Department of Energy Science and Engineering, IIT Delhi. His group currently focuses on exploring the fundamental and applied aspects of metal-sulfur batteries.

Contributors

Vipin Kumar
Department of Energy Science and Engineering, Indian Institute of Technology Delhi, University of Queensland–IIT Delhi Academy of Research (UQIDAR), Indian Institute of Technology Delhi, Hauz Khas, New Delhi, India

C Sanjaykumar
School of Interdisciplinary Research, Indian Institute of Technology Delhi, Hauz Khas, New Delhi, India

Rahul Singh
Department of Energy Science and Engineering, Indian Institute of Technology Delhi, Hauz Khas, New Delhi, India

Rajendra Singh
Department of Physics, Indian Institute of Technology Delhi, Hauz Khas, New Delhi, India

Chhail Bihari Soni
Department of Energy Science and Engineering, Indian Institute of Technology Delhi, Hauz Khas, New Delhi, India

Sungjemmenla[1]
Department of Energy Science and Engineering, Indian Institute of Technology Delhi, Hauz Khas, New Delhi, India

SK Vineeth
University of Queensland–IIT Delhi Academy of Research (UQIDAR), Indian Institute of Technology Delhi, Hauz Khas, New Delhi, India

Yusuke Yamauchi
University of Queensland–IIT Delhi Academy of Research (UQIDAR), Indian Institute of Technology Delhi, Hauz Khas, New Delhi, India
Australian Institute for Bioengineering and Nanotechnology (AIBN), The University of Queensland, Brisbane, QLD 4072, Australia

---

[1] Author does not have any surname

1 Introduction
Sodium-sulfur battery technology

S.K. Vineeth and Vipin Kumar

1.1 INTRODUCTION: SODIUM-SULFUR BATTERY TECHNOLOGY

Energy has become indispensable as it is a global commodity and a key element regulating worldwide development [1]. With the relevance of clean, sustainable energy sources and environmental considerations, renewable energy sources have received the utmost research importance [2]. The intermittence of renewable energy sources, however, decelerated its wide acceptance. The quest for an effective solution to the abovementioned issue catalyzed the growth of energy storage technology [3,4]. The very concept of energy storage technology can be observed from nature. The fundamental process known as photosynthesis is a basic example of energy conversion and storage. Photosynthesis involves converting solar energy to chemical energy and subsequent storage in the plant cell, thereby generating food for its survival. From the storage viewpoint, the energy must be stored in a highly reversible form that does not cause any dissipation upon extraction.

In particular, electrochemical energy storage devices such as batteries have been identified as potential solutions for stationary and portable applications [5]. With their intercalation chemistry, lithium-ion batteries (LIBs) have reached a specific energy of ~260 Wh·kg^{-1}, close to its theoretical limit [6,7]. The limited availability of raw materials and concerns regarding price inflation and the environment resulted in the expedition for battery technologies beyond LIBs [4,8]. However, sodium, the fourth most abundant element on earth, with a redox potential of –2.7 V vs. SHE (standard hydrogen electrode), possesses a high specific discharge capacity of 1165 mAh g^{-1} [9]. When paired with a high-capacity cathode, sodium produces a high-energy density battery. One such technology is metal-sulfur batteries, where sodium-sulfur (Na-S) battery technology has a prominent role [10]. From a material perspective, the Na-S battery employs metallic sodium (Na) as anode material and a composite sulfur consisting of sulfur as active material (with/without electrocatalyst), a conductive filler, and a binder as a cathode [11]. An electrolyte is also employed, which maintains internal contact with the electrode systems. High energy density, lower cost per kW h, availability of raw materials, and cost-effective nature are the factors that uprise Na-S batteries compared to Li-ion technology.

Furthermore, considering the cost parameters, sodium accounts for approximately 4% of that of Li [12]. At the cathode part, sulfur being the active material is a petrochemical byproduct with a high theoretical specific capacity of 1672 mAh g^{-1} on a two-electron reaction [13]. Sulfur is the 17th most abundant element on earth, with an elemental abundance of sulfur accounting for approximately 953 ppm in the upper continental crust [14]. Hence, from the energy and natural abundance point of view, the overall cost per kW h for Na-S batteries could be way lower than that of LIBs. Therefore, Na-S battery chemistry could be a potential candidate for the post-LIBs era.

1.2 BRIEF HISTORY OF NA-S BATTERY

Early developments in Na-S batteries focused on high-temperature sodium-sulfur (HT Na-S) batteries operated at 300°C or above. The chemistry leads to a high energy density of about 760 Wh kg^{-1}. The suitability of HT Na-S batteries for electric vehicles was examined by Kummer et al. in 1966, which employed a ceramic solid electrolyte β″-Al$_2$O$_3$ to isolate molten sodium and sulfur [15,16]. From the 1980s to the 1990s, the applicability of HT Na-S batteries was also tested for aerospace applications [17]. Followed by this, NGK Insulators, Ltd and the Tokyo Electric Power Company (TEPCO) collaborated and developed commercial-level Na-S batteries in early 2003 [18]. In particular, a commercial-level 34-MW HT Na-S battery cluster was established in 2009 in Japan to stabilize a 51 MW wind farm [19]. However, safety aspects decelerated the widespread adoption of HT Na-S batteries. In this aspect, intermediate room temperature Na-S (IM Na-S) batteries, which operate at relatively lower temperatures of 120–300°C with a similar sulfur redox reaction as that of HT Na-S batteries, were developed by Abraham et al. in the 1970s [20,21]. From the viewpoint of the materials, the IM Na-S battery consists of sodium polysulfides (Na$_2$S$_n$) dissolved in a non-aqueous catholyte and liquid sodium metal as the anode. A ceramic electrolyte was employed as the electrolyte. Due to lower temperature operation, the maintenance cost could be lowered dramatically, which is one of the attractive features that favored IM Na-S battery development.

In 2006, the development of a polymer-based electrolyte comprising polyethylene oxide (PEO) electrolyte operated at 90°C was reported by Park et al. [22]. A year later, a room-temperature sodium-sulfur (RT Na-S) battery with liquid electrolyte was demonstrated by Wang et al. [23]. Though Na-S chemistry could be operated at ambient temperature, identifying polysulfides, dissolution, and cross-over (i.e., shuttling) in electrolytes was found accountable for their limited cycle life [24]. While dendrite propagation and unstable solid-electrolyte interface (SEI) creation were the issues associated with the anode, the capacity fade was linked with sluggish kinetics, the poor electrical conductivity of sulfur particles, and active material loss due to the dissolution of polysulfides. To enhance the charge transfer kinetics, attempts were made to design a cathode host with an electrocatalyst to accelerate sluggish kinetics. Electrocatalysis, e.g., atomic cobalt, was employed to boost the performance of RT Na-S batteries [25]. Significant efforts have been made in stabilizing

Introduction: Na-S battery technology

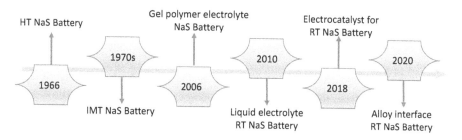

FIGURE 1.1 Timeline of the development of sodium-sulfur batteries.

electrode-electrolyte interfaces through electrolyte engineering and anode modifications. Sodium-alloy and biphasic (dual-components) interphases were notable developments in electrode engineering for RT Na-S batteries [26,27]. Figure 1.1 presents a timeline of the development of Na-S batteries.

1.3 SODIUM METAL BATTERIES AND THE OPERATION OF HT NA-S BATTERIES

Presently, two types of sodium metal batteries exist; HT Na-S batteries (commercially comes in the form of tubular cells) and zero-emission battery research association (ZEBRA), both relying on the liquid nature of the sodium metal anode and sulfur cathode [28]. These battery chemistries are preferred to their peers, e.g., Pb-acid batteries, due to their high energy and long cycle life, which drives down the cost per cycle. Figure 1.2 depicts the development of sodium-based batteries along with theoretically/practically predicted/obtained energy densities, respectively.

Since both the chemistries are operated at a high temperature (i.e., 300°C–350°C), the molten electrodes, i.e., sodium (melting point ~98°C at 1 atm) and sulfur (melting point ~113°C at 1 atm) or metal halide ($NiCl_2$), facilitate the reversible and stable operation of the cell without significant raise in the polarization loses at the interface [29]. The engineers of Ford Motors first developed HT Na-S battery technology in the 1960s, followed by design modifications by the National Aeronautics and Space Administration (NASA) [24]. The Tokyo Electric Power Company developed and commercialized modern Na-S batteries in collaboration with NGK insulators in Japan. In the past two decades, HT Na-S batteries have been deployed globally to offer attractive, cost-competitive technology for large-scale electrical energy storage [30]. HT Na-S battery modules are economically viable only if installed with a minimum installed capacity of MW [31]. Owing to high theoretical specific capacity and high solubility, the HT Na-S system can deliver a high theoretical energy density of 760 Wh/kg, many-fold to that of Pb-acid. Though the practical energy density varies from 180 to 220 Wh/kg or 990 to 121 Wh per 5.5 kg of Na-S, it is much higher than Pb-acid batteries and most Li-ion chemistries.

If one notices critically, the Na-S cell construction is the same as for a conventional flooded Pb-acid battery except for high-temperature operation and the nature of the electrolyte. The structure of both batteries is depicted schematically in Figure 1.3.

Due to their solid electrolyte and liquid electrodes, the Na-S chemistry is a bit unusual compared to conventional rechargeable batteries, where electrodes and electrolytes are in the solid and liquid states. The temperature decides the nature of discharge products. For instance, the lower valent discharge products that contribute to capacity build-up are achievable at relatively higher temperatures (over 600°C). Though a high operating temperature helps to increase the overall energy density, it significantly improves the operation cost and lowers the overall energy efficiency.

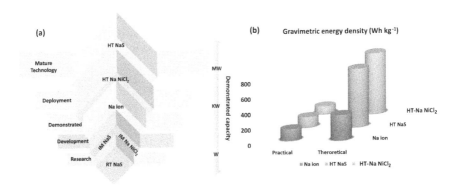

FIGURE 1.2 (a) Current development status of the leading Na-based battery technologies and their (b) Theoretically predicted or practically obtained specific energy densities.

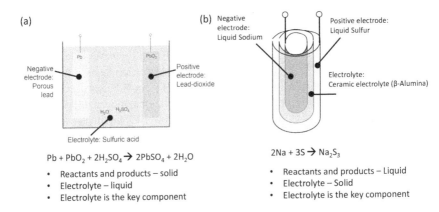

FIGURE 1.3 Schematic representation showing the design aspects of (a) lead acid battery. (b) HT Na-S battery.

Introduction: Na-S battery technology

In addition, a high-temperature operation affects the cycle life. Generally, Na-S chemistry is operated at 300°C–400°C, which helps increase the reversibility while minimizing precipitation of the solid-discharge products. The reaction is stopped when it reaches a composition of Na_2S_3 instead of Na_2S_2 or other lower valent discharge products.

1.4 THE TRANSITION FROM HIGH-TEMPERATURE TO RT NA-S BATTERIES

Generally, the discharge curve of an HT Na-S battery exhibits two voltage plateaus. The voltage plateau formed at the higher cell voltage (above 2.0 V) suggests the formation of higher-order polysulfides. HT Na-S chemistry creates various polysulfide species; Na_2S_5 is the first polysulfide formed at a higher voltage. It is to be noted that Na_2S_5 and sulfur are immiscible, and hence the cell voltage decays faster. A voltage plateau at approximately 2.08 V vs. Na/Na^+ suggests the phase change [28]. The cell voltage continuously decreases during the transformation from Na_2S_5 to Na_2S_4 and to Na_2S_3, which follows a single phase and reaches a voltage of 1.74 V. Further reduction in voltage below 1.74 V leads to the formation of Na_2S_2 and Na_2S, which are solids and may deposit over inorganic electrolyte surface or at cathode [29]. Hence, the discharge potential is restricted to 1.74 V to ensure the reversibility of the cell. Though it improves the reversibility, it limits the system to reach a capacity of 1672 mAh g^{-1} (considering Na_2S as a final product). It restricts the maximum achievable capacity to 557 mAh g^{-1} (based on Na_2S_3 as the final product) [15]. Moreover, a higher operating temperature hinders the stoichiometric window of Na to S (~0.66), limiting the energy density of the system. The electrochemical reaction regarding HT Na-S cell is shown below:

Anode: $2Na \rightarrow 2Na^+ + 2e^-$

Cathode: $x\,S + 2e^- \rightarrow 2S_x^{2-}$

Overall cell reaction: $2Na^+ + x\,S \rightarrow Na_2S_x$

It is to be highlighted that the polysulfide melts formed are corrosive to electrodes, current collectors, and even the casing materials of the cell/battery. Hence judicious selection needs to be made. A schematic of the cell architecture of HT Na-S with its cell reaction is shown in Figure 1.4.

A lower operating temperature of Na-S chemistry is the key to its higher energy density, improved safety, and widespread adoption; however, it is highly challenging to operate Na-S chemistry at ambient conditions reversibly. One such challenge is poor sulfur utilization, affecting cell capacity. The BASE electrolyte's room-temperature or intermediate temperature (<200°C) ionic conductivity is insufficient to support the electrochemical process. Therefore, other ceramic electrolytes (e.g., NASICON) were explored to meet the requirements [32]. The Na^+-ion conducting ceramic electrolytes could show promising results but are not sufficient to consider for deployment.

To allow Na-S chemistry to operate reversibly at a relatively lower temperature (preferably below 200°C), Abraham et al. and NASA tweaked the electrochemistry

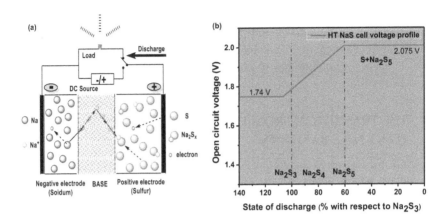

FIGURE 1.4 (a) Schematic showing the interfacial architecture of HT Na-S cell. (b) A typical voltage profile of a HT Na-S cell with phases existing at each separate phase of the state of charge.

of the system [24]. The Na-S battery operating at a lower temperature is termed an intermediate-temperature Na-S battery (IM Na-S). The IM Na-S exhibits similar reaction pathways to HT Na-S chemistry; Na_2S_2 is the final discharge product instead of Na_2S_3 [33]. As a result, IM Na-S batteries could attain a relatively higher discharge capacity than HT Na-S batteries. Several studies considering different issues of IM Na-S batteries have been considered in the past two decades [34]. The most recent addition is the design of a cathode comprising a sodium polysulfide solution dissolved in a suitable organic solvent with nanoscale carbon. Lowering the operating temperature invites various inevitable issues. For instance, the solubility of sodium polysulfide is compromised in organic solvents causing poor utilization of sulfur. The ionic conductivity at lower temperature cases increased polarization losses. Moreover, the amount of sodium required to operate the cell is much higher than the high-temperature operation due to the lower transference number of sulfide ions. Moreover, the undissolved polysulfide precipitate promotes blockage at the electrolyte surface, thereby hindering ion migration.

The development of organic electrolytes and their successful employment in the Na-S system made the realization of achieving RT Na-S battery [35]. Electrochemically, there is a vast difference in the electrochemical reaction for HT Na-S, IM Na-S, and RT Na-S chemistry. In the case of HT and IM Na-S chemistry, the maximum achievable energy density is about 760 Wh kg^{-1}, and Na_2S_3 and Na_2S_2 are considered the final discharge products. However, RT Na-S could reversibly transform elemental sulfur to Na_2S, following various stages of transition from sulfur to polysulfides. Table 1.1 shows an overview of HT, IM, and RT Na-S battery systems and their cell components. Mainly, four transition regions are of great interest. The

TABLE 1.1
Overview of the transition from HT Na-S to RT Na-S chemistries

Battery system	Developmental stage	Nature of anode	Nature of cathode	Nature of electrolyte	Operational temperature and cell design	Achievable capacity
HT Na-S	Commercialized but discarded due to safety issues	Molten sodium	Molten sulfur	Solid electrolyte	300°C–350°C	557 mAh g^{-1}
IM Na-S	Developed but not commercialized	Molten sodium	Molten sulfur/ semi-solid	Solid electrolyte	120°C–300°C	557 mAh g^{-1}
RT Na-S	Research level	Solid sodium modified with interphases	Solid sulfur in the form of a composite sulfur cathode system	Liquid, quasi-solid, and solid-state electrolytes	25°C	~1200 mAh g^{-1}

reaction chemistries listed below show each transition stage and the corresponding products formed:

1. Solid-liquid transition region
   $S_8 + 2Na^+ + 2e^- \rightarrow Na_2S_8$ ~2.20 V vs. Na/Na$^+$
2. Liquid-liquid transition region
   $Na_2S_8 + 2Na^+ + 2e^- \rightarrow 2Na_2S_4$ 2.20–1.65 V vs. Na/Na$^+$
3. Liquid-solid transition region
   $Na_2S_4 + 2/3\ Na^+ + 2/3e^- \rightarrow 4/3\ Na_2S_3$ ~1.65 V vs. Na/Na$^+$
   $Na_2S_4 + 2Na^+ + 2e^- \rightarrow 2Na_2S_2$
   $Na_2S_4 + 6Na^+ + 6e^- \rightarrow 4Na_2S$
4. Solid-solid transition region
   $Na_2S_2 + 2Na^+ + 2e^- \rightarrow 2Na_2S$ 1.65–1.20 V vs. Na/Na$^+$

Compared to HT and IM Na-S batteries, RT Na-S chemistry could reversibly produce Na$_2$S as the final discharge product, achieving a high theoretical energy density of ~1274 Wh kg^{-1}. Figure 1.5 shows a typical discharge profile for the IM Na-S and RT Na-S batteries. The ambient temperature operation could reduce the safety and maintenance concerns required for the HT or IM Na-S battery systems. However, various other challenges popped out in the case of RT Na-S batteries.

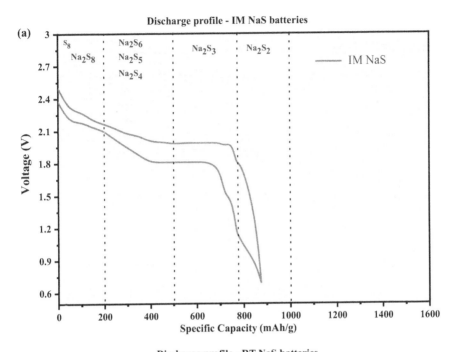

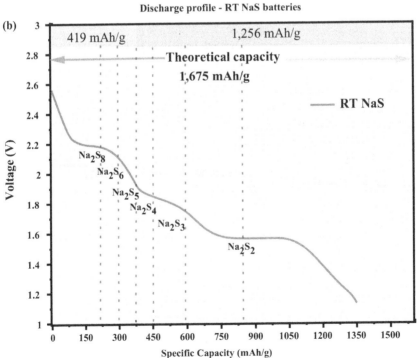

FIGURE 1.5 A typical discharge profiles for (a) IM Na-S battery. (b) RT Na-S battery.

1.5 DEVELOPMENT OF THE RT NA-S BATTERIES

Numerous attempts were made in early 2010 to develop liquid electrolytes and examine their compatibility with sodium metal anodes and composite sulfur cathodes [22,36,37]. The battery system comprises a sulfur cathode, i.e., a sulfur-infused carbon-based cathode, and a sodium metal anode separated by a polyolefin separator. A liquid electrolyte was used to conduct the ions. Most of the trials failed after a few cycles, and it was credited to the polysulfide dissolution, shuttling of polysulfides, and subsequent anode poisoning [35].

Since sulfur is highly insulating, the RT Na-S battery employs a carbon-mixed sulfur cathode instead to ensure improved electronic conductivity. The choice of nanoscale carbon is vital in ensuring the long-term cycling stability of the RT Na-S batteries. For example, a micro-mesoporous carbon host promotes stability for a few hundred cycles, while a microporous carbon leads to a much longer cycle life. The micropores of a microporous carbon physically trap the discharge products and restrict them from entering the electrolyte. However, a lower sulfur loading in microporous carbon limits their use.

RT Na-S battery favors conversion reactions; therefore, the volume change upon charge/discharge reactions is a matter of concern. Since the volume change in the cathode is significant, a binder is generally used to ensure the integrity of the cathode. The sulfur cathode is formulated as a composite material with a conductive matrix, typically carbon-based material and binder which amalgamates carbon particles and sulfur molecules. Binders are usually polymer molecules that can adhere to particles, ensuring particle-to-particle contact and contact with the current collector [38–44]. The main attractive feature of polymeric binders is that they can be solution-processed, and by solvent evaporation, a continuous film can be formed [45–51]. It is noteworthy that selecting a suitable binder regulates the electrochemical performance of the RT Na-S battery. For instance, Alex et al. showed the importance of cathode-binder interaction by evaluating the effects of various binders for RT Na-S batteries [52]. It was demonstrated that binders with polar groups, such as carboxyl groups, interacted with short- and long-chain polysulfides and showed no cracks indicating their ability to accommodate volume expansion with cycling.

Introducing carbon materials can positively enhance charge transfer kinetics at the cathode-electrolyte interface. High surface area and high electrical conductivity regulate sulfur utilization, stabilizing RT Na-S system performance [53]. The carbon amount is lower than the active material in the general fabrication procedure. A high amount of carbon can hinder ionic conduction by forming separate aggregates and can lead to phase separation of cathode slurry. Generally, the slurry composition has the following constituents: sulfur is about 65–70 wt.%, conducting carbon (Carbon black or Ketjan black) ~15–20 wt.%, and polymeric binder (i.e., PVDF) ~10 wt.%. The solvent employed is generally organic solvent, such as *N*-Methylpyrrolidone; however, it could be aqueous depending upon the choice of the polymer binder (e.g., CMC or sodium alginate). The solid particles are ball milled before adding solvent for uniform mixing and enhancing the effective surface area.

The loading of active materials regulates the RT Na-S battery performance. Pope and Aksay reported the correlation between sulfur loading and energy density

[54]. For achieving ~400 Wh/kg of specific energy density, active material loading is estimated at 2 mg/cm^2 [55]. Hence tailoring cathodes for high-performance RT Na-S systems requires an overall high sulfur loading in the matrix. Though a high sulfur loading is necessary, the reported sulfur loadings vary from <1 mg/cm^2 to >5 mg/cm^2 [15]. The main reason for not increasing sulfur loading is the negative effect of polysulfide dissolution and shuttling, which grows exponentially with sulfur loading [11]. Besides high sulfur loading, the liquid electrolyte was found to be equally responsible for deteriorating battery performance [56]. To mitigate this issue, the ratio of electrolyte volume to the sulfur loading (E/S) was adjusted to improve the performance. For instance, an E/S ratio of 4–5 μl/mg is found suitable to extend the cycling stability of the cell. Generally, the range reported in the literature for a coin cell varies from 50 to 100 μl/mg [57].

Though sodium metal as an anode relishes a high theoretical specific capacity of 1166 mAh g^{-1} with an electrochemical potential of –2.71 V, the commercialization of RT Na-S batteries is being hampered due to the instabilities of the sodium metal anode. Sodium metal is highly reactive toward most organic electrolytes; the electrolyte gets reduced on the surface of the anode. The reduced species lead to forming of a thick, fractured, and mechanically weak solid electrolyte interphase (SEI) on the sodium metal anode [58]. The sodium metal anode is prone to fail upon cycling due to unwarranted growth of dendrites, inevitable side reactions, dissolution in the liquid electrolyte, etc. The research on anode modifications mainly aims to alleviate the abovementioned challenges. Although the exact reason for dendrite growth has not been divulged, it can be closely correlated with the uneven ion flux generated due to non-uniformity in the electric field distribution. Facilitating a uniform deposition of sodium ions through interfacial engineering is a promising approach to assuage dendrite growth.

Regarding the stabilization of SEI, one widely explored technique is to engineer electrolytes and employ artificial solid-state electrolytes [59]. The volume expansion has been controlled through the introduction of a sodium host. However, in-depth knowledge of the root causes and conditions must be precisely tracked to protect sodium effectively. Table 1.2 gives an overview of various challenges associated with each component of the RT Na-S battery and its remedies.

In the case of electrolytes for RT Na-S batteries, factors including ionic conductivity, transference number, and electrochemical stability window (ESW) determine the cell performance. ESW of an electrolyte is defined as the potential range within which the electrolyte neither undergoes reduction nor oxidation. Electrolytes with wider ESW are favorable for any battery chemistry. ESW is related to the energy levels of solvent molecules with respect to the anode and cathode [60]. In addition, physical and thermal properties such as wettability with separator, viscosity, and flash point also need to be considered while selecting an electrolyte. For liquid electrolytes, the properties of solvents also function as a determinant factor for selection criteria [61]. The solvent should have a high dielectric constant, high polarity, moderate Lewis's acidity/basicity, and lower viscosity values. Strong Lewis's basicity of solvent endorses better solvation of sodium salt and promotes ionic mobility. Another parameter that regulates electrolyte properties is the solvation [62]. Understanding solvation structure helps to correlate SEI formation and battery performance [63].

Introduction: Na-S battery technology

To enhance the performance of the RT Na-S battery, the electrode-electrolyte interface (EEI) is engineered through various routes. One of the widely used methods is the introduction of electrolyte additives [64]. Electrolyte additives is a broad terminology that includes additives that form alloy interface with sodium metal [26,65], SEI stabilizing additives [66], cathode-electrolyte interphase (CEI) stabilizing additives [67], additives for controlling dendrite growth [68], flame retardant additives [69], etc. Ionic mobility through the EEI is determined by solvation-desolvation dynamics [70]. Molecular interactions are the basics of the concept of solvation [71]. Interactions between ions and solvent molecules alter the viscosity of electrolyte and ionic conductivity [72]. Experimental and theoretical calculations are employed together to determine the solvation properties of electrolytes [73].

Liquid electrolytes possess high ionic conductivity and better electrode-electrolyte wettability and are widely reported for RT Na-S batteries. However, they are volatile, with a low flash point causing safety concerns. Hence considering safety as a prime concern, alternatives for liquid electrolytes have also been explored [74].

TABLE 1.2
A summary of the significant challenges associated with each component of RT Na-S battery and their corresponding remedies

Battery component	Challenges	Remedies
Anode	1. Dendrite issues 2. Unstable SEI 3. Volume expansion 4. Gas evolution and safety	• Interfacial engineering, interlayers, and alloy interface creation • Electrolyte additives, quasi-solid state, and solid-state electrolytes • Anode host • Electrodeposited anodes
Cathode	1. Insulating nature of sulfur and its discharge species 2. Polysulfide dissolution and shuttling 3. Sluggish kinetics 4. Volume expansion	• The composite cathode consists of conducting carbon and sulfur. • Physical confinement, introducing blocking layers and ion-selective membranes • Developing an engineered host for accommodating sulfur • Sulfur-infused cathode systems, melt diffused and infiltrated cathode composite • Elemental doping and electrocatalyst-incorporated cathodes
Electrolyte	1. Unstable SEI formation 2. Polysulfide dissolution and shuttling 3. Flammability and safety aspects	• Electrolyte additives employing advanced electrolytes such as solid-state electrolytes • Formulating high-concentrated electrolytes, and localized high-concentrated electrolyte systems • Modified separators and ion-selective membranes • Incorporating flame retardants and functional fillers in polymer electrolytes • Tailoring quasi-solid state and solid-state electrolytes

The liquid electrolyte also has the limitation of a tendency to form unstable SEI. Polymer electrolytes are considered a better option as they are mechanically stable, resist dendrites, and are suitable for large-scale applications [75]. Quasi-solid state and solid-state polymer electrolytes are significant subdivisions [76,77]. In addition, inorganic solid electrolytes have also been employed for RT Na-S batteries [24,78]. However, poor ionic conduction and high interfacial resistance are the factors that limit its wide acceptance [79]. Nevertheless, considering safety aspects, polymer, and solid-state electrolytes have immense potential, and the research scope in this field has been booming for the past few years [80].

1.6 CONCLUSION AND PROSPECTS

In the quest for a high energy density battery system, Na-S chemistry has been a fascinating technology. Though initial developments focussed on designing and scaling HT Na-S batteries, which showed considerable success, safety concerns decelerated the growth. Commercialized HT Na-S batteries were developed in Japan, which could supply energy demands. However, the great interest in HT Na-S technology plummeted as safety concerns and strong policies were imposed. It should be noted the thermal runaway leading to an explosion in Japan made the HT Na-S technology be discarded at the commercial level. Alternatives to the HT Na-S system paved the way for IM Na-S technology which again had limitations and did not gain much appreciation. A notable milestone in the second half of the 2000s in employing liquid and polymer electrolytes for Na-S batteries revealed a newer technology, RT Na-S batteries. Although the cells showed significant capacity decay and short cycling life, tailored sodium metal anode and composite sulfur cathodes showed extended cycling stability.

To sum up, from the transition of HT Na-S to IM Na-S and finally to RT Na-S technology, there have been significant efforts concerning the design, architecture, and materials perspective. However, at each stage, specific bottleneck challenges could be identified. Timely rectifications have been done by unveiling the root cause of each. Advanced characterization techniques and in situ analysis have a definite role in unveiling the complexities. It can be profoundly stated that the amalgamation of computational and experimental procedures can shed more light in revealing the mechanisms which can catalyze the realization of a commercial, high-performance, stable RT Na-S battery for high-energy applications.

REFERENCES

1. Sungjemmenla, Vineeth SK, Soni CB, et al. Understanding the Cathode–Electrolyte Interphase in Lithium-Ion Batteries. *Energy Technol.* 2022;10:2200421.
2. Soni CB, Arya N, Sungjemmenla, et al. Microarchitectures of Carbon Nanotubes for Reversible Na Plating/Stripping Toward the Development of Room-Temperature Na–S Batteries. *Energy Technol.* 2022;10(12):2200742.
3. Sundaram PM, Soni CB, Sungjemmenla, et al. Reviving Bipolar Construction to Design and Develop High-Energy Sodium-Ion Batteries. *J Energy Storage.* 2023;63:107139.
4. Chandra M, Shukla R, Saroha R, et al. Physical Properties and Electrochemical Performance of Zn-Substituted $Na_{0.44}Mn_{1-x}Zn_xO_2$ nanostructures as cathode in Na-ion batteries. *Ceram Int.* 2018;44:21127–21131.

5. Chandra M, Khan TS, Shukla R, et al. Diffusion Coefficient and Electrochemical Performance of NaVO$_3$ anode in Li/Na batteries. *Electrochim Acta*. 2020;331:135293.
6. Jishnu NS, Vineeth SK, Das A, et al. Electrospun PVdF and PVdF-co-HFP-Based Blend Polymer Electrolytes for Lithium Ion Batteries. In: Balakrishnan NTM, Prasanth R, editors. *Electrospinning for Advanced Energy Storage Applications*. 1st ed. Singapore: Springer Singapore; 2021. pp. 201–234.
7. Nitta N, Wu F, Lee JT, et al. Li-ion Battery Materials: Present and Future. *Mater Today*. 2015;18:252–264.
8. Saroha R, Khan TS, Chandra M, et al. Electrochemical Properties of Na$_{0.66}$V$_4$O$_{10}$ Nanostructures as Cathode Material in Rechargeable Batteries for Energy Storage Applications. *ACS Omega*. 2019;4:9878–9888.
9. Hwang J-Y, Myung S-T, Sun Y-K. Sodium-ion Batteries: Present and Future. *Chem Soc Rev*. 2017;46:3529–3614.
10. Soni CB, Sungjemmenla, Vineeth SK, et al. Patterned Interlayer Enables a Highly Stable and Reversible Sodium Metal Anode for Sodium-Metal Batteries. *Sustain Energy Fuels*. 2023;7:1908–1915.
11. Sungjemmenla, Soni CB, Vineeth SK, et al. Unveiling the Physiochemical Aspects of the Matrix in Improving Sulfur-Loading for Room-Temperature Sodium–Sulfur Batteries. *Mater Adv*. 2021;2:4165–4189.
12. Abraham KM. How Comparable Are Sodium-Ion Batteries to Lithium-Ion Counterparts? *ACS Energy Lett*. 2020;5:3544–3547.
13. Sungjemmenla, Soni CB, Vineeth SK, et al. Exploration of the Unique Structural Chemistry of Sulfur Cathode for High-Energy Rechargeable Beyond-Li Batteries. *Adv Energy Sustain Res*. 2022;3:2100157.
14. Hans Wedepohl K. The Composition of the Continental Crust. *Geochim Cosmochim Acta*. 1995;59:1217–1232.
15. Wang Y, Zhou D, Palomares V, et al. Revitalising Sodium–Sulfur Batteries for Non-High-Temperature Operation: A Crucial Review. *Energy Environ Sci*. 2020;13:3848–3879.
16. Kummer JT, Weber N. Battery Having a Molten Alkali Metal Anode and a Molten Sulfur Cathode. *US Patent*. 1968;3413150A.
17. Chang R, Minck R. Sodium-Sulfur Battery Flight Experiment Definition Study. *J Power Sources*. 1990;29:555–563.
18. Oshima T, Kajita M, Okuno A. Development of Sodium-Sulfur Batteries. *Int J Appl Ceram Technol*. 2005;1:269–276.
19. Kawakami N, Iijima Y, Fukuhara M, et al. Development and Field Experiences of Stabilization System Using 34MW NAS Batteries for a 51MW Wind Farm. *2010 IEEE Int Symp Ind Electron*. 2010;2371–2376.
20. Abraham KM, Rauh RD, Brummer SB. A Low Temperature Na-S Battery Incorporating A Soluble S Cathode. *Electrochim Acta*. 1978;23:501–507.
21. Fielder WL, Singer J. Solubility, Solubility, Stability, and Electrochemical Studies of Sulfur-Sulfide Solutions in Organic Solvents. *NASA Tech Pap*. Number is No. NASA-TP-1245. 1978;1–40.
22. Park C, Ryu H, Kim K, et al. Discharge Properties of All-Solid Sodium–Sulfur Battery Using Poly (Ethylene Oxide) Electrolyte. *J Power Sources*. 2007;165:450–454.
23. Wang J, Yang J, Nuli Y, et al. Room Temperature Na/S Batteries with Sulfur Composite Cathode Materials. *Electrochem Commun*. 2007;9:31–34.
24. Vineeth SK, Tebyetekerwa M, Liu H, et al. Progress in the Development of Solid-State Electrolytes for Reversible Room-Temperature Sodium–Sulfur Batteries. *Mater Adv*. 2022;3:6415–6440.
25. Zhang B-W, Sheng T, Liu Y-D, et al. Atomic Cobalt as an Efficient Electrocatalyst in Sulfur Cathodes for Superior Room-Temperature Sodium-Sulfur Batteries. *Nat Commun*. 2018;9:4082.

26. Kumar V, Eng AYS, Wang Y, et al. An Artificial Metal-Alloy Interphase for High-Rate and Long-Life Sodium–Sulfur Batteries. *Energy Storage Mater.* 2020;29:1–8.
27. Kumar V, Wang Y, Eng AYS, et al. A Biphasic Interphase Design Enabling High Performance in Room Temperature Sodium-Sulfur Batteries. *Cell Reports Phys Sci.* 2020;1:100044.
28. Hueso KB, Armand M, Rojo T. High Temperature Sodium Batteries: Status, Challenges and Future Trends. *Energy Environ Sci.* 2013;6:734.
29. Hueso KB, Palomares V, Armand M, et al. Challenges and Perspectives on High and Intermediate-Temperature Sodium Batteries. *Nano Res.* 2017;10:4082–4114.
30. Wen Z, Cao J, Gu Z, et al. Research on Sodium Sulfur Battery for Energy Storage. *Solid State Ionics.* 2008;179:1697–1701.
31. Andriollo M, Benato R, Dambone Sessa S, et al. Energy Intensive Electrochemical Storage in Italy: 34.8MW Sodium–Sulphur Secondary Cells. *J Energy Storage.* 2016;5:146–155.
32. Anantharamulu N, Koteswara Rao K, Rambabu G, et al. A Wide-Ranging Review on Nasicon Type Materials. *J Mater Sci.* 2011;46:2821–2837.
32. Lu X, Kirby BW, Xu W, et al. Advanced Intermediate-Temperature Na–S Battery. *Energy Environ Sci.* 2013;6:299–306.
34. Nikiforidis G, van de Sanden MCM, Tsampas MN. High and Intermediate Temperature Sodium–Sulfur Batteries for Energy Storage: Development, Challenges and Perspectives. *RSC Adv.* 2019;9:5649–5673.
35. Li T, Xu J, Wang C, et al. The Latest Advances in the Critical Factors (Positive Electrode, Electrolytes, Separators) for Sodium-Sulfur Battery. *J Alloys Compd.* 2019;792:797–817.
36. Ryu H, Kim T, Kim K, et al. Discharge Reaction Mechanism of Room-Temperature Sodium-Sulfur Battery with Tetra Ethylene Glycol Dimethyl Ether Liquid Electrolyte. *J Power Sources.* 2011;196:5186–5190.
37. Kim I, Kim CH, Choi SH, et al. A Singular Flexible Cathode for Room Temperature Sodium/Sulfur Battery. *J Power Sources.* 2016;307:31–37.
38. Vineeth SK, Gadhave RV. Corn Starch Blended Polyvinyl Alcohol Adhesive Chemically Modified by Crosslinking and Its Applicability as Polyvinyl Acetate Wood Adhesive. *Polym Bull.* 2023. https://doi.org/10.1007/s00289-023-04746-0.
39. Gadhave RV, Vineeth SK, Mahanwar PA, et al. Effect of Addition of Boric Acid on Thermo-Mechanical Properties of Microcrystalline Cellulose/Polyvinyl Alcohol Blend and Applicability as Wood Adhesive. *J Adhes Sci Technol.* 2021;35:1072–1086.
40. Vineeth SK, Gadhave RV, Gadekar PT. Investigation of Crosslinking Ability of Sodium Metabisulphite with Polyvinyl Alcohol–Corn Starch Blend and Its Applicability as Wood Adhesive. *Indian Chem Eng.* 2022;64:197–207.
41. Gadhave RV, Vineeth SK. Synthesis and Characterization of Starch Stabilized Polyvinyl Acetate-Acrylic Acid Copolymer-Based Wood Adhesive. *Polym Bull.* 2022. https://doi.org/10.1007/s00289-022-04558-8.
42. Vineeth SK, Gadhave RV, Gadekar PT. Polyvinyl Alcohol–Cellulose Blend Wood Adhesive Modified by Citric Acid and Its Effect on Physical, Thermal, Mechanical and Performance Properties. *Polym Bull.* 2022. https://doi.org/10.1007/s00289-022-04439-0.
43. Dhawale PV, Vineeth SK, Gadhave RV, et al. Tannin as a Renewable Raw Material for Adhesive Applications: A Review. *Mater Adv.* 2022;3:3365–3388.
44. Gadhave RV, Vineeth SK, Dhawale PV, et al. Effect of Boric Acid on Poly Vinyl Alcohol-tannin Blend and Its Application as Water-Based Wood Adhesive. *Des Monomers Polym.* 2020;23:188–196.
45. Vineeth SK, Gadhave RV, Gadekar PT. Chemical Modification of Nanocellulose in Wood Adhesive: Review. *Open J Polym Chem.* 2019;09:86–99.

46. Vineeth SK, Gadhave RV, Gadekar PT. Glyoxal Cross-Linked Polyvinyl Alcohol-Microcrystalline Cellulose Blend as a Wood Adhesive with Enhanced Mechanical, Thermal and Performance Properties. *Mater Int.* 2020;2:0277–0285.
47. Singh HK, Patil T, Vineeth SK, et al. Isolation of Microcrystalline Cellulose from Corn Stover with Emphasis on Its Constituents: Corn Cover and Corn Cob. *Mater Today Proc.* 2020;27:589–594.
48. Vineeth SK, Gadhave RV, Gadekar PT. Nanocellulose Applications in Wood Adhesives—Review. *Open J Polym Chem.* 2019;09:63–75.
49. Vineeth SK, Gadhave RV. Sustainable Raw Materials in Hot Melt Adhesives: A Review. *Open J Polym Chem.* 2020;10:49–65.
50. Gadhave RV, Vineeth SK, Gadekar PT. Cross-Linking of Polyvinyl Alcohol/Starch Blends by Glutaraldehyde Sodium Bisulfite for Improvement in Thermal and Mechanical Properties. *J Mater Environ Sci.* 2020;11:704–712.
51. Dhawale PV, Vineeth SK, Gadhave RV, et al. Cellulose Stabilized Polyvinyl Acetate Emulsion: Review. *Open J Org Polym Mater.* 2021;11:51–66.
52. Eng AYS, Nguyen D-T, Kumar V, et al. Tailoring Binder–Cathode Interactions for Long-Life Room-Temperature Sodium–Sulfur Batteries. *J Mater Chem A.* 2020;8:22983–22997.
53. Kumar D, Mishra K. A Brief Overview of Room Temperature Na-S Batteries Using Composite Sulfur Cathode. *Macromol Symp.* 2021;398:1900206.
54. Pope MA, Aksay IA. Structural Design of Cathodes for Li-S Batteries. *Adv Energy Mater.* 2015;5:1500124.
55. Liu Y, Liu S, Li G-R, et al. High Volumetric Energy Density Sulfur Cathode with Heavy and Catalytic Metal Oxide Host for Lithium–Sulfur Battery. *Adv Sci.* 2020;7:1903693.
56. Mu P, Dong T, Jiang H, et al. Crucial Challenges and Recent Optimization Progress of Metal–Sulfur Battery Electrolytes. *Energy Fuels.* 2021;35:1966–1988.
57. Syali MS, Kumar D, Mishra K, et al. Recent Advances in Electrolytes for Room-Temperature Sodium-Sulfur Batteries: A Review. *Energy Storage Mater.* 2020;31:352–372.
58. Soni CB, Sungjemmenla, Vineeth SK, et al. Challenges in Regulating Interfacial-Chemistry of the Sodium-Metal Anode for Room-Temperature Sodium-Sulfur Batteries. *Energy Storage.* 2022;4:e264.
59. Wang Y-X, Zhang B, Lai W, et al. Room-Temperature Sodium-Sulfur Batteries: A Comprehensive Review on Research Progress and Cell Chemistry. *Adv Energy Mater.* 2017;7:1602829.
60. Chen X, Shen X, Hou T-Z, et al. Ion-Solvent Chemistry-Inspired Cation-Additive Strategy to Stabilize Electrolytes for Sodium-Metal Batteries. *Chem.* 2020;6:2242–2256.
61. Xu X, Zhou D, Qin X, et al. A Room-Temperature Sodium–Sulfur Battery with High Capacity and Stable Cycling Performance. *Nat Commun.* 2018;9:3870.
62. Tian Z, Zou Y, Liu G, et al. Electrolyte Solvation Structure Design for Sodium Ion Batteries. *Adv Sci.* 2022;9:2201207.
63. Vineeth SK, Soni CB, Sun Y, et al. Implications of Na-Ion Solvation on Na Anode–Electrolyte Interphase. *Trends Chem.* 2022;4:48–59.
64. Wang H, Wang C, Matios E, et al. Facile Stabilization of the Sodium Metal Anode with Additives: Unexpected Key Role of Sodium Polysulfide and Adverse Effect of Sodium Nitrate. *Angew Chemie Int Ed.* 2018;57:7734–7737.
65. Zheng X, Fu H, Hu C, et al. Toward a Stable Sodium Metal Anode in Carbonate Electrolyte: A Compact, Inorganic Alloy Interface. *J Phys Chem Lett.* 2019;10:707–714.
66. Zhao X, Zhu Q, Xu S, et al. Fluoroethylene Carbonate as an Additive in a Carbonates-Based Electrolyte for Enhancing the Specific Capacity of Room-Temperature Sodium-Sulfur Cell. *J Electroanal Chem.* 2019;832:392–398.

67. Moeez I, Susanto D, Chang W, et al. Artificial Cathode Electrolyte Interphase by Functional Additives toward Long-Life Sodium-Ion Batteries. *Chem Eng J.* 2021;425:130547.
68. Kreissl JJA, Langsdorf D, Tkachenko BA, et al. Incorporating Diamondoids as Electrolyte Additive in the Sodium Metal Anode to Mitigate Dendrite Growth. *ChemSusChem.* 2020;13:2661–2670.
69. Feng J, An Y, Ci L, et al. Nonflammable Electrolyte for Safer Non-Aqueous Sodium Batteries. *J Mater Chem A.* 2015;3:14539–14544.
70. Monti D, Jónsson E, Boschin A, et al. Towards Standard Electrolytes for Sodium-Ion Batteries: Physical Properties, Ion Solvation and In-Pairing in Alkyl Carbonate Solvents. *Phys Chem Chem Phys.* 2020;22:22768–22777.
71. Li Y, Lu Y, Adelhelm P, et al. Intercalation Chemistry of Graphite: Alkali Metal Ions and Beyond. *Chem Soc Rev.* 2019;48:4655–4687.
72. Andreev M, de Pablo JJ, Chremos A, et al. Influence of Ion Solvation on the Properties of Electrolyte Solutions. *J Phys Chem B.* 2018;122:4029–4034.
73. Eng AYS, Soni CB, Lum Y, et al. Theory-Guided Experimental Design in Battery Materials Research. *Sci Adv.* 2022;8. https://doi.org/10.1126/sciadv.abm2422
74. Eng AYS, Kumar V, Zhang Y, et al. Room-Temperature Sodium–Sulfur Batteries and Beyond: Realizing Practical High Energy Systems through Anode, Cathode, and Electrolyte Engineering. *Adv Energy Mater.* 2021;11:2003493.
75. Lopez J, Mackanic DG, Cui Y, et al. Designing Polymers for Advanced Battery Chemistries. *Nat Rev Mater.* 2019;4:312–330.
76. Lei D, He YB, Huang H, et al. Cross-Linked Beta Alumina Nanowires with Compact Gel Polymer Electrolyte Coating for Ultra-Stable Sodium Metal Battery. *Nat Commun.* 2019;10:1–11.
77. Zhou D, Chen Y, Li B, et al. A Stable Quasi-Solid-State Sodium–Sulfur Battery. *Angew Chemie Int Ed.* 2018;57:10168–10172.
78. Hayashi A, Noi K, Sakuda A, et al. Superionic Glass-Ceramic Electrolytes for Room-Temperature Rechargeable Sodium Batteries. *Nat Commun.* 2012;3:856.
79. Famprikis T, Canepa P, Dawson JA, et al. Fundamentals of Inorganic Solid-State Electrolytes for Batteries. *Nat Mater.* 2019;18:1278–1291.
80. Yang J, Zhang H, Zhou Q, et al. Safety-Enhanced Polymer Electrolytes for Sodium Batteries: Recent Progress and Perspectives. *ACS Appl Mater Interfaces.* 2019;11:17109–17127.

2 Sodium metal anode
Past, present, and future of sodium metal anode

Chhail Bihari Soni and Vipin Kumar

2.1 INTRODUCTION TO THE SODIUM METAL ANODE

Advanced battery technologies with high specific energies are required to address the rising demands of grid storage and electrification in the future. To this end, we must constantly innovate and create better battery chemistries to store more significant amounts of energy in a given space, see Figure 2.1a. Owing to relatively high theoretical specific capacity (~1166 mAh g^{-1}) and fairly positive redox potential (0.34 V vs. Li metal), research interest in sodium-based batteries has increased many folds in the last decade [1–3], see Figure 2.1b. The performance of the existing Li-ion batteries can easily be leap-frogged by rechargeable sodium metal batteries [4]. Sodium-based battery technologies, including room-temperature sodium-sulfur batteries (RT Na-S) [5], RT-Na-oxygen batteries [6], and RT-ZEBRA batteries [7], have the technological and economic interest due to the low-cost and high energy of these chemistries [8].

Moreover, a high natural abundance of sodium metal brings new merits over Li-based batteries. Sodium is abundant across the region [9]; see Figure 2.1c, which avoids monopoly in the supply chain of the required materials. For example, about 23 billion tons of soda ash (Na_2CO_3), one of the main sources of sodium, is identified in the United State alone. Sodium, one of life's vital components, had long been with us; however, the metal could be isolated through electrolysis of sodium hydroxide, which Sir Humphry Davy first reported in 1807 [10]. Several other methods, including brine, rock mining, and seawater evaporation, can also obtain sodium metal.

Due to sodium's relatively positive standard reduction potential, it does not undergo alloying reaction with the battery components, unlike Li-based batteries where Li undergoes alloying reaction with the current collectors; copper or aluminum can be used as current collectors for sodium-based batteries [11]. When paired with a sulfur cathode, alkali metals, such as sodium, lead to very high-energy battery chemistry [12–14], see Figure 2.1d.

Unlike Li metal anode, Na metal anode reacts with most organic electrolytes more vigorously, forming a thick, broken interphase layer that contains majorities of the reduced organic species [15,16]. The interphase is found to be mechanically and chemically unstable, causing the sodium metal to react repeatedly with the organic

DOI: 10.1201/9781003388067-2

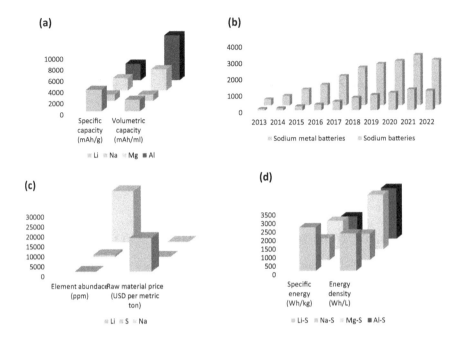

FIGURE 2.1 (a) Specific and volumetric capacities of lithium, sodium, magnesium, and aluminum metal anodes. (b) Number of publications over the years for sodium metal batteries and sodium batteries. Data was collected from the Web of Science portal, with keywords "sodium metal battery" and sodium "battery." (c) Element abundance and raw material price for Li. S and Na. (d) specific energy and energy density for Li-S, Na-S, Mg-S, and Al-S batteries.

electrolyte. The stability of the sodium metal anode is significantly impacted by the ionic conduction of sodium ions across the interphase [17], Na$^+$ ions flux, nucleation and growth of sodium dendrites, sodium deposition uniformity, the electrochemical potential of organic molecules, and thermodynamic integrity [18].

The non-uniform deposition of Na$^+$ ions causes the localized perturbation in the physical and chemical properties of the interface (such as surface energy, stress and defects, and self-decomposition), eventually leading to the formation of dendrites [8]. In addition, during the plating/stripping processes, the Na metal anode experiences extremely large volume change. The very nature of the Na metal anodes' failure and degradation provides the essential understanding necessary to demonstrate the reversible operation of the anode. An internal short and subsequent thermal runaway of the battery comes from Na deposition on the Na metal anode, often exhibiting a dendritic morphology, which is shown to penetrate the separator film and reach the cathode.

The significance of the Na metal anode had been grossly undervalued. However, in recent years, the importance of Na metal anodes has come to light, and many inventive efforts have offered potential methods to deal with Na metal's problems.

2.2 CHALLENGES IN DEVELOPING A STABLE SODIUM METAL ANODE

It is crucial to prepare the Na metal strips (a thin slice of a thickness of about 100 μm) before assembling the Na metal-based batteries. Na metal strips are not offered in stores like Li metal. Pure Na metal must be stored in mineral oil or an inert gas atmosphere. If exposed to the ambient, it can quickly produce an oxide/hydroxide passivation layer.

Besides that, numerous fundamental challenges require immediate attention to deploy Na metal batteries successfully. Figure 2.2 presents a birds-eye view of the significant challenges that are accountable for hampering the development of Na metal batteries.

2.2.1 Unstable solid electrolyte interphase

As described briefly in the last section, the SEI formed on the sodium metal anode is inherently unstable, causing limited cycle life and the Coulombic efficiency [8,19]. Though the electrochemical and thermodynamic properties of the Na metal differ a bit from the Li metal, it is sufficient to cause a significant change in the formation of the SEI [20]. It is noteworthy to mention that the stability of the SEI on the surface of the sodium metal anode is strongly dependent on the chemical environment of the electrolyte. For example, in most carbonate-based electrolytes, e.g., EC-DMC, EC-PC, PC, or EC-EMC, the SEI is predicted to be thick, and its thickness increases with cycle life. The localized chemical information of the SEI suggests the formation

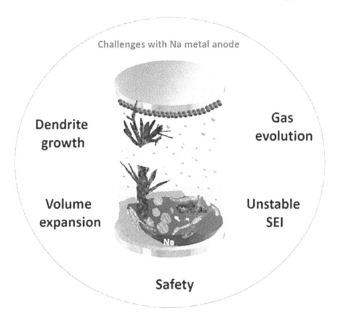

FIGURE 2.2 Schematic illustration of the major issues associated with the sodium metal anode.

of organic-rich SEI in the environment of carbonate electrolytes [17]. However, a thin, compact, and inorganic-rich SEI is often observed to form in the environment of ether-based electrolytes [8]. Despite significant progress in forming a stable SEI, the performance is yet to be improved to ensure the long-term stability of the sodium metal anode.

2.2.2 Dendrite Growth

The dendrite growth is one of the critical issues that decides the stability of the sodium metal anode. Though the exact cause of dendrite growth is not yet known for sodium metal anodes, it is believed that the localized inhomogeneities lead to the development of sodium dendrites. Unlike lithium dendrites, sodium dendrites are mechanically weaker and prone to detach from the metal surface. There is a dearth of relevant studies on sodium dendrite growth. A few studies consider the flaws and cracks in the SEI as the root cause of the dendrite growth, while others assume that a localized non-uniformity in the current density leads to sodium dendrites' growth [21,22]. According to Chazalviel [23], if the concentration of ions close to the metal anode falls to zero, a space charge layer is created that causes a strong electric field to surround the electrode, accelerating the nucleation process and thereby, growth of Na dendrites. Irrespective of the root cause of the sodium dendrite growth, a non-uniform Na deposition is often achieved. A non-uniform deposition is not suitable for the stability of the cell, as it causes a rise in the reaction overpotential, decay in the Coulombic efficiency, and an increase in cell resistance [19]. It is previously reported that tiny protrusions and irregular metal ion flux delivery may promote a non-uniform sodium deposition [24]. Many theoretical and experimental strategies have been put forth to reduce dendritic growth and improve the health of the SEI [25–28]. Recent attempts have identified that a flat and atomically smooth metal surface achieves a thin, compact, and electrochemically stable interphase.

In addition, it has been observed that the dendrites have a self-amplifying feature that causes them to grow until a dead filament (i.e., one disconnected from the electrode surface) is created [29]. This has been empirically verified by in situ microscopic methods [30]. Furthermore, when the dendrite grows, a new, highly reactive metal becomes accessible for the electrolyte to react with, increasing the electrode overpotential, directly related to the development of the SEI. Sand's model allows us to connect the behavior of dendritic growth with the applied current density [24,31].

Hong's team conducted an in situ experiment using a quasi-zero electrochemical field to track the growth pattern of Na metal dendrites [32]. The preferred growth was ascribed to the non-uniform Na^+ ion deposition on the metal electrode surface during a plating cycle. The energy barriers for nucleating the Na metal on the fresh metal surface were significantly higher than on the non-uniform surface. The electrode was exposed to a nearly zero electrochemical field to observe the effect of electrolytes on the formation of dendrites. Due to limited mechanical strength, Na dendrites shrink and dissolve into the electrolyte. The results of the trials showed that the growth of Na dendrite follows the prediction of Sand's model. It is clear from the literature that

Sodium metal anode

the fundamental understanding of the sodium dendrites' nucleation and growth is lousy, likely due to a lack of suitable analytical techniques.

2.2.3 Instabilities due to localized physical changes

There are various known and unknown reasons for the localized physical instabilities of the sodium metal anode; the volume change during the stripping/plating process substantially impacts the well-being of the cell. Since virtually infinite sodium atoms are available at the metal anode side, a limitless volume expansion and contraction occurs during repeated cycling [33]. The volume change causes sodium metal anode and electrolyte loss, leading to Coulombic efficiency loss [33,34]. The occurrence of volume change is thought to relieve internal stresses of the Na metal anode due to dendrite growth [33]. The mechanical failure of the Na metal anode, which includes cracks, pulverization, dissolving, and structure collapse, is often caused by volume changes [35–38].

The interfacial energy could be one of the critical parameters in deciding the physical stability of the sodium metal anode. For example, it has been found that the instabilities in the sodium metal anode can be minimized through alloy phases that facilitate the wettability of the electrolyte [39–41]. Generally, alloy reactions favor a decrease in the reaction's Gibbs free energy, the only factor that can reduce interfacial instabilities. More recently, Zhu et al. [42] identified that alkali-earth metals like Be, Mg, and Ba could essentially abolish the physical instabilities of the SEI. Because these metals are readily miscible with Na metal (i.e., tend to form an alloy with Na metal) and allow the creation of a buffer layer of solid solution to minimize structural instabilities in the SEI, they may help lower the nucleation barriers for Na deposits. Besides that, the nucleation overpotential and particle coarsening are connected to the wetting properties of the anode. It has been noted that the nucleation potential for Na or Li over Ni and Cu is slightly higher (>50 mV), which is thought to be caused by an incompatibility between the crystal structure and lattice parameters (i.e., Ni and Cu belong to the face central cubic (FCC) crystal structure with 1.24 and 1.28 atomic radii. In contrast, Na and Li belong to the body-centered cubic (BCC) structure with 4.28 and 1.52 atomic radii, respectively) [43].

In addition, the gas evolution contributes to the localized physical instabilities of the sodium metal anode [44,45]. The Na metal reacts vigorously with most organic solvents. It may cause the evolution of carbon monoxide, propane, and other hydrocarbons, making the Na metal anode vulnerable to pressure build-up within the cell [46–48]. It is important to note that the chemistry of the salt-solvent, or ion-solvent complexes, significantly impacts the gas generation rate [44]. Due to the solid mechanical seal and thick casing walls, gas evolution in coin cells is typically undetectable. The gas evolution issue, however, significantly impacts battery performance for large-area batteries like pouch cells.

The relative rates of the gas evolution in various solvent systems, such as DEC/FEC, FEC/PC, and DEC/EC in the presence of the Na metal anode, are monitored by Mullin's group [45]. The DEC, PC, or EC alone was shown to have the highest gas evolution rates. They observed that the presence of FEC in the electrolyte could help partially prevent gas evolution.

A thorough understanding of the interaction between Na⁺ ions and the substrate at the atomic and molecular levels is required to rationalize Na metal anodes.

2.2.4 Instability due to localized chemical changes

The localized chemical changes at the anode surface play a vital role in deciding the stability of the sodium metal anode. The composition of the interphase is dynamic, i.e., it changes with cycle number, and so do its properties. For instance, an organic-rich interphase often leads to decay in the cycling stability of the sodium metal anode. Such interphases barely protect the sodium metal anode against parasitic side reactions and repeated electrolyte consumption. Conversely, an interphase rich in inorganic materials supports extended cycle longevity while preserving the electrolyte's integrity. Nonetheless, the elimination of organic constituents from the SEI remains a formidable task. Though attempts are made to calculate the composition of the SEI, it isn't straightforward to determine the true composition of the SEI precisely. The uncertainty in the chemical composition of the SEI often leads to false predictions about the stability of the metal anode [49]. Like other battery anodes, Na metal batteries' organic and inorganic components (such as $ROCO_2Na$, HC OONa, polyethylene oxide oligomers, and NaF, NaCl, and Na_2CO_3, respectively) are primarily derived from the reduction of the organic and inorganic components of the electrolyte [48]. The chemical makeup of both components significantly influences the stability of the SEI. For instance, an SEI with a high inorganic content inhibits dendritic formation, but its low flexibility makes it vulnerable to mechanical alterations [50]. However, the SEI with many organic components can withstand substantial mechanical changes, whereas the SEI with a low mechanical strength cannot effectively control dendritic formation. Therefore, the SEI's chemical stability is determined by providing an optimal composition of both components. Through altered electrolyte systems, attempts have been made to modify the SEI's functionality.

More recently, it has been discovered that the SEI derived from ether-based solvents with $NaN(SO_2CF_3)_2$ salt facilitates the production of both components in well-optimized ratios [51]. Both components were found to be distributed in the horizontal rather than the vertical plane of the SEI despite having well-optimized compositions (i.e., across the thickness). For example, the interior of the SEI had a higher concentration of inorganic components (such as NaF and Na_2O). Still, the upper surface of the SEI had a higher concentration of organic ingredients.

Like ether-based solvents, carbonate-based solvents (such as EC/DMC and EC/DEC) react strongly with Na metal and cause the surface to generate a highly uncontrolled SEI. The SEI in carbonate-based solvents is predicted to be fragile and mechanically inferior to the ether-based electrolytes.

2.3 STRATEGIES TO OVERCOME THE CHALLENGES

The issues mentioned above are the major roadblock in developing a stable and long-life sodium metal anode. In light of the above, constant efforts are in progress to stabilize the sodium metal anode. The following are major strategies developed to stabilize the sodium metal anode: (1) Surface polishing, (2) fabrication of the intrinsic interphase, (3) fabrication of the artificial interphase, (4) Nanostructured current collector, and (5) guided deposition.

2.3.1 SURFACE POLISHING

Attempts are made to stabilize the SEI to suppress the dendrite growth. However, it is arduous to ensure the long-term stability of the SEI, or it is next to impossible to develop an ideal SEI. The SEI, which is chemically and physically uniform, compact, thin, permeable to sodium ions without affecting the integrity of the electrolyte, and electrically insulating, is considered an idea for long-term battery operation.

In the quest to develop an ideal SEI, Gu et al. reported an electrochemical surface polishing approach to fabricate atomically flat and smooth SEI on the alkali metal anodes by manipulating electrolyte reduction processes [52]. Electrochemical surface polishing is a technique to remove the metal selectively from the surface of the electrode under the application of an electric field. The electrolyte plays a vital role in controlling the flatness of the surface. The stripping/plating of the alkali metal anodes was performed at the potential well below their oxidation potential, and therefore, the choice of the electrolyte became essential. For example, a smooth SEI on the Li metal anode was realized using DOL-DME-LiTFSI in a potential window of 0.6-1.4 V vs. Li/Li$^+$. A thin SEI of thickness of about 12 nm with a smooth surface of roughness of about 1.5 nm was confirmed by atomic force microscopy (AFM). The surface polishing approach is a highly scalable process to achieve a desired surface profile by controlling the applied electric field.

2.3.2 ELECTROLYTE ENGINEERING

In RT Na-based batteries, liquid electrolytes are frequently employed. To design a stable sodium anode, it is essential to tweak the interactions between the sodium metal anode and the organic molecules of the liquid electrolytes. Generally, due to the relative LUMO position of most organic molecules, they get reduced and adsorbed on the surface of the metal anode. The residue serves as the SEI, which is fractured and mechanically unstable. The nature of the SEI can be tuned by tuning the local chemistry of the electrolyte. The choice of the electrolyte solvent and salts can alter the local health of the SEI.

2.3.2.1 Electrolyte solvent and salts for a stable sodium metal anode

In the higher LUMO position, the ether-based electrolytes preferentially induce salt decomposition rather than solvent decomposition, resulting in an inorganic-rich SEI with a high Na$^+$ conductivity.

Earlier attempts were made to study 1 M sodium hexafluorophosphate (NaPF$_6$) in bis(2-methoxyethyl) ether to create Na anodes with excellent Coulombic efficiency and longer-cycle life [53]. The surface of the Na anode was shown to have a consistent SEI made of inorganic sodium fluoride (NaF) and sodium oxide (Na$_2$O). A thin and ionically conductive SEI is often formed with NaPF$_6$ salt. However, the salt is observed to promote side reactions if not handled properly. A fractional amount (above 100 ppm) of the moisture leads to generate hydrogen fluoride (HF), which corrodes the battery components and affects the overall battery performance. Wang et al. [54] recently reported a cutting-edge electrolyte containing $NaBF_4$ salt and diglyme solvent that achieved an unprecedentedly high average CE of 99.93% over 400 cycles for stable Na metal anodes. The high ion conductivity B-O species act as a strong binder to tightly connect the inorganic components in the inner SEI part and the organic polymer in the top layer, resulting in a continuous and flexible SEI layer. Notably, Na salt also plays a vital role in creating SEI. The stability of the SEI is identified to decrease in diglyme electrolytes when the salt anion is changed from OTf > PF_6 > ClO_4 > FSI > TFSI. It is evident from the choice of the anion that selecting a suitable electrolyte salt is critical to stabilizing the sodium metal anode.

2.3.2.2 Electrolyte additives

The local chemistry of the electrolytes can be tuned with the help of the additives. The additive salt or solvent must be soluble in the mother solvent. The additive salts can be of various types, for instance, allying and solid electrolyte types. The alloying type additive salts are the one that gets reduced at the sodium metal anode and facilitates the formation of alloying interphase [55]. The examples include but are not limited to $SnCl_2$ and InI_3. The solid electrolyte-type additives favor the formation of an interphase that behaves similarly to a solid electrolyte. The classic example of such additives includes P_2S_5 and Na_2S or a combination of both.

However, the additive solvent could be of different types, for example, reducing or chemically inert. The reducing type additive solvents are preferably reduced on the surface of the sodium metal anode before the electrolyte solvent to form a stable SEI. Example of such additives includes but are not limited to fluoroethylene carbonate (FEC). The inert-type additive solvents are either poor or unable to solvate the electrolyte salt and hence unable to participate in the solvation process; however, due to molecular crowding, the properties of the SEI change significantly. For instance, tetrafluoroethyl tetrafluoropropyl ether (TTE) serves as the inert type solvent additive to alter the local chemistry of the electrolyte.

Any change in the chemistry of the electrolyte directly affects the properties of the SEI. The additives enable the formation of a thin and compact SEI with a homogenous chemical and physical structure.

More recently, a new type of additive salt is examined, i.e., potassium bis(trifluoromethyesulfonyle)imide (KTFSI), which serves as a bi-functional electrolyte additive [56]. It is suggested that K^+ ions preferentially get adsorbed on the sodium metal surface and act as electrostatic shielding to mitigate dendritic growth. However, TFSI⁻ anion decomposes into oxynitrides, and sodium nitrate that offers an ideal N-containing SEI. This cation-anion coordination provides a highly stable Na metal anode.

Considering the drawbacks of ether solvents, such as their high cost, high volatility, and poor oxidation potential, it is essential and necessary to produce carbonate electrolytes for Na metal anodes. Additionally, to increase the safety of the electrolyte, flame-retardant additives such as perfluoro-2-methyl-3-pentanone [57] and 1, 1, 1, 3, 3, 3-hexafluoroisopropanol methyl ether [58] can be added. More research is required to advance the additives used for sodium metal batteries.

2.3.2.3 High-concentration electrolytes

High-concentration electrolytes (HCEs) are a novel class of electrolytes where the salt concentration is higher than 3 M. Generally, HCEs have a higher ion transference number, a wider electrochemical window compared to their lean counterpart, higher thermal stability, a higher flash point, and good flame resistance. Thanks to their distinct solvation topologies, the HCEs demonstrated high stability in contact with the sodium metal anodes. The HCEs can enable a stable and reversible sodium metal anode. Their large-scale applications may encounter difficulties due to their high cost, high viscosity, and poor wettability. Attempts are made to overcome the issues mentioned above. The HCEs are subjected to an additional solvent phase to ensure a decrease in the viscosity and improve the wettability. The solvent must be "inert," i.e., not participate in the solvation events. Though the "inert" solvent does not partake in the solvation process, it promotes the crowding of the solvated molecules. This arrangement is often called a localized high-concentration electrolyte (LHCE). Hydrofluoro ethers, an "inert" solvent with a low donor number and dielectric constant, had little to no impact on the HCE's initial solvent structure [59]. The LHCE demonstrated excellent stability for Na anodes with reduced viscosity and improved wettability while maintaining a high Na-ion transference number. Table 2.1 provides a comprehensive picture of the liquid electrolyte systems used hitherto for sodium metal batteries.

2.3.3 Interface engineering – designing intrinsic and extrinsic interphases

Manipulating the interface between the anode and electrolyte affects the stability of the sodium metal anode. When the electrode/electrolyte interface is altered through electrolyte additives, such manipulation is called "intrinsic interphase." The intrinsic interphases are thin, compact, and grow inherently, improving performance significantly. However, the intrinsic interphases are subject to non-uniformity, side reactions, and dissolution with cycle life. A definite redox relationship must exist between the sodium metal anode and the electrolyte additives to realize the formation of intrinsic interphases. An example of intrinsic interphases may include the surface of the sodium metal anode in the presence of $SnCl_2$ and InI_3 electrolyte additives that often gets reduced on the surface of the sodium metal anode to form a compact and thin layer to protect the sodium metal anode. The intrinsic interphases can effectively inhibit the growth of dendrites and boost the stability of the sodium metal anode.

However, when the surface of the metal anode can be treated with an additional solid or liquid phase to grow a thin layer, the approach is called "extrinsic interphase." The extrinsic interphases are easy to control and independent of the redox properties of the elements. Therefore, a wide variety of materials can be used to form extrinsic interphases. For instance, the reaction between sodium metal anode and 1-bromopropane leads to a thin coating of NaBr, demonstrating a low diffusion barrier for interfacial ion transport. The compact NaBr interphase avoids unfavorable side interactions between the electrode and electrolyte and restricts dendritic growth.

TABLE 2.1
A comparative assessment of electrolyte engineering to stabilize the Na metal anode

Na cell	Salt	Solvent		Approaches/Technologies/Methods to stabilize Na metal anode	Current density (mA/cm^2)	Capacity (mA h/cm^2)	Cycle life (h)	Pro's	Con's
Na-Na	1 M NaOTf	TEGDME	Electrolyte engineering	0.01 M KTFSI [56]	0.5	1	1200	Stable SEI, non-dendritic deposition	Bulky TFSI anions affect the charge transfer resistance
Na-Na	1 M NaPF$_6$	DIglyme		CTAB as electrolyte additive [60]	10	30	400	Synergistic effect of CTA$^+$ cation and Br$^-$ anion to enhance the stability and electrochemical performance of a Na metal anode	Sophisticated characterization tools are needed to investigate the effect
Na-Na	4 M NaFSI	DME		1% SbS$_3$ [41]	0.5	0.5	1000	Na ion rapid transportation, effective suppression of dendrites, and stable cycling	High electrolyte concentration raises the concern about thermal stability and cost
Na-Cu	1 M NaPF$_6$	Diglyme		[53]	0.5	1	1200	Highly reversible, non-dendritic Na plating/stripping, uniform inorganic SEI highly permeable to electrolyte solvent	Though NaPF$_6$ is beneficial with ether-based solvents, it does not help Na metal anode's reversibility in carbonate-based electrolytes
	1 M NaPF$_6$	Monoglyme			0.5	1	1200		
	1 M NaPF$_6$	EC/DEC			0.5	1	100		

(Continued)

TABLE 2.1 (CONTINUED)

Na cell	Salt	Solvent	Approaches/Technologies/Methods to stabilize Na metal anode	Current density (mA/cm²)	Capacity (mA h/cm²)	Cycle life (h)	Pro's	Con's
Na-Na	1 M NaClO₄	EC/DEC	0.05 M SnCl₂ as additive [39]	0.5	1	500	Fast interfacial ion transfer passivates electrolyte with Na anode to mitigate reactivity, non-dendritic deposition	The deposition morphology is highly rough and coarse, which leads to raising the overpotential
Na-Na	1 M NaPF₆	EC/DMC	3% FEC [61]	0.05	-	600	Growth of passivation layer that hinders the high reactivity of Na towards electrolyte	High deposition/dissolution overpotential (200 mV)
Na-Na	2.1 M NaFSI	DME	BTFE (Bis(2,2,2-trifluoroethyl) ether) as an additive [59]	1	1	>950	Dendrite-free Na deposition with high CE (>99%)	High viscosity, poor wettability, and high additive cost
Na-Na	1 M NaPF₆	PC/EC	FEC (Fluoroethylene carbonate) as an additive [62]	1	1	100	Effective suppression of Na dendrites, homogeneous morphology of SEI, and mechanically stable SEI	The Coulombic efficiency (CE) is low (~88%), which is likely due to the irreversible formation of fluoro-compound of sodium

Electrolyte engineering

For a uniform and conformal deposition of the extrinsic interphase, atomic layer deposition (ALD) and molecular layer deposition (MLD) techniques are used [63,64]. To deposit an incredibly thin layer of Al_2O_3 on sodium metal, a low-temperature plasma-enhanced ALD (PE-ALD) method is adopted [65]. Besides that, 2D materials are also used to design the extrinsic interphase on the sodium metal anode. Studies suggest that 2D materials with atomically thin and uniform defects benefit ionic conductivity, lowering diffusion barriers and improving Na-ion diffusion [66]. However, the stiffness and hardness of the 2D materials are found undesirable for inhibiting the dendrites.

Tables 2.2 and 2.3 comprehensively summarize the materials systems used to design extrinsic interphases on the sodium metal anode.

2.3.4 Nanostructured Host and Current Collector

The sodium metal is soft and susceptible to fracture upon repeated charge/discharge cycles. Therefore, it is essential to design a host for the anode to ensure the mechanical stability of the sodium metal anode. The host material must be electrochemical neutral and must not interfere with the redox reactions [75]. An ideal host material for sodium metal anode must possess: (a) more positive redox potential with respect to sodium metal, (b) large internal spacing or porosity to accommodate volume change, (c) excellent wettability (or sodiophilicity), (d) good mechanical properties to prevent dendrite formation, and (e) low cost and ease of fabrication for practical applications. The carbon-based materials satisfy most requirements to serve as the host for alkali metal anodes.

A processable and moldable Na metal anode is designed by contacting reduced-graphene oxide (rGO) on molten sodium metal. The capillary effect pulls molten sodium between the inter-layer gaps [76]. This design was inspired by the sodiophilic features of reduced graphene oxide (rGO).

The approach to designing a host appears to have several benefits, including volume change, dendrites growth due to homogenous deposition, and increased safety by acting as a buffer to adapt to changing stress levels in the interlayer region. Table 2.4 summarizes the materials systems examined hitherto for the sodium metal anodes.

2.4 FUTURE PROSPECTS

Even though there has been advancement in developing a stable and reversible sodium metal anode, there is still room for further improvement, especially when scaling up the process to the pouch cell.

Expecting a breakthrough in the upcoming year is, at best, misguided and, at worst, arrogant. But when looking at the long-term potential of sodium batteries, it is clear that Na metal-based batteries will quickly become essential in energy storage, which is a necessity in the modern world. Fortunately, there is a lot of optimism for further development. In terms of advanced characterization and protective measures, the relatively mature studies on Li metal anodes conducted over the past 40 years have helped guide the protection of Na metal. As a result, Na metal-based batteries

TABLE 2.2
A comparative assessment of layered and coating material to stabilize the Na metal anode

Na cell	Salt	Solvent	Approaches/Technologies/Methods to stabilize Na metal anode	Current density (mA/cm^2)	Capacity (mA h/cm^2)	Cycle life (h)	Pro's	Con's
Na-Na	1 M NaPF$_6$	Diglyme	Nano-SiO$_2$ as coating material [67]	1	1	800	Homogeneous distribution of Na ion flux, effective suppression of Na dendrite	Not suitable with carbonate-based electrolyte
Na-Na	1 M NaPF$_6$	EC/PC	Na$_3$PS$_4$ as coating material [68]	1	1	270	Less parasitic side reactions, uniform SEI, and improved Na ion flux	A large deposition or dissolution overpotential (300 mV)
Na-Na	1 M NaPF$_6$	EC/PC	PhS$_2$Na$_2$- rich layer [69]	5	1	280	Effectively prevents the growth of dendrites, highly stable Na metal electrodeposition.	Large overpotential (400 mV)
Na-Na	1 M NaPF$_6$	EC/PC	MLD Alucone [63]	1	1	270	Non-dendritic deposition, enhanced lifetime, better performance than ALD Al$_2$O$_3$	A significant deposition or dissolution overpotential (200 mV)
Na-Na	1 M NaClO$_4$ 1 M NaCF$_3$SO$_3$	PC DGME	Carbon paper as coating material with 5% FEC as an additive [70]	1 5	1 1	400 480	Enhance electrochemical performance even at significant current densities, favorable SEI due to less reactive surface of carbon paper	A significant deposition of dissolution overpotential (60 mV)
Na-Na	1 M NaClO$_4$	EC/PC	Ionic membrane [71]	0.1	—	250	Reduced Na metal reactivity without compromising transportation of ions	High cost of ionic liquids, less current density

(Continued)

TABLE 2.2 (CONTINUED)

Na cell	Salt	Solvent	Approaches/Technologies/Methods to stabilize Na metal anode	Current density (mA/cm^2)	Capacity (mA h/cm^2)	Cycle life (h)	Pro's	Con's
Na-Na	1 M NaPF$_6$	EC/DEC	Graphene as coating material [25]	0.5	0.5	200	Dendrite-free Na metal anode, highly stable	The viability of the method is challenging. Graphene layer thickness limits the performance at a high current density
Na-Na	1 M NaPF$_6$	EC/PC	NaBr coating [72]	0.25	0.25	500	Fewer energy barriers to the transport of ions that is comparable to metallic magnesium	Lack of mechanical stabilities (NaBr is brittle) affects the stability of the interphase
Na-Na	1 M NaCF$_3$SO$_4$	DEGDME	ALD Al$_2$O$_3$ [64]	3	1	500	Remarkably increased Na plating/stripping performance, and non-dendritic growth	Sophisticated and slow deposition process, lack of flexibility limits its ability to accommodate volume change
Na-Na	1 M NaClO$_4$	EC/DEC	Plasma enhanced ALD Al$_2$O$_3$ [65]	0.25	1	400	Growth of the passivation layer and remarkable increase in cyclic performance	Sophisticated and slow deposition process, lack of flexibility limits its ability to accommodate volume change
Na-Na	1 M NaClO$_4$	EC/PC	Inorganic-organic compositive protective layer [26]	0.5	1	>550	Reduced electrolyte decomposition, enhanced reversibility	A trade-off between ionic conductivity and share modulus of FCPL

TABLE 2.3
A comparative assessment of composite and alloy materials to stabilize the Na metal anode

Na cell	Salt	Solvent	Approaches/Technologies/Methods to stabilize Na metal anode	Current density (mA/cm^2)	Capacity (mA h/cm^2)	Cycle life (h)	Pro's	Con's
Na/Na	1 M NaPF6	EC-DEC-FEC	Artificial interlayer composed of NaBr/Na$_3$P nano-crystallines (NaBrP) [73]	1	1	700	Enhanced electrochemical kinetics and interfacial stability of Na metal anode from a spontaneous reaction between Na metal and PBr$_3$ solutions	Compromised cycle life at a high current density
MAI/Na-MAI/Na	1 M NaCF$_3$SO$_3$	DEGDME	Artificial metal alloy interphase [55]	2	1	650	Reversible deposition, even at high current density (2–7 mA/cm^2), mechanically suppression of dendrite growth	A significant deposition or dissolution overpotential (700 mV) at high current density (7 mA/cm^2)

(Composite/alloy Material)

(*Continued*)

TABLE 2.3 (CONTINUED)

Na cell	Salt	Solvent	Approaches/Technologies/Methods to stabilize Na metal anode	Current density (mA/cm^2)	Capacity (mA h/cm^2)	Cycle life (h)	Pro's	Con's
BH/Na- BH/Na	1 M NaPF$_6$	EC/DEC/PC + 5% FEC	Composite Na electrode with NaSICON type Na ion conductor (BH-Na) (Bulk hybrid Na metal electrode, SnO$_2$@NZSP) [50]	1	5	700	Accommodate the volume change and enhance mechanical strength to hinder the growth of dendrites	Deposition or dissolution overpotential varies with cycles (50–400 mV), which predominantly affects the deposition morphology and the Coulombic efficiency
Na/Sb@ZMF/C-Na/Sb@ZMF/C	1 M NaPF$_6$	Diglyme	Pre-sodiated alloy-induced interphase (carbon-coated zinc+ encapsulated Sb nanocrystals) [51]	1	1	800	Stable Na plating/stripping	A significant deposition or dissolution overpotential
Na/NHC-Na/NHC	1 M NaCF$_3$SO$_3$	diglyme	N-functionalized hard carbon [74]	1	1	1700	Mechanical suppression of dendrite growth, excellent cyclic stability	Inferior performance at a high current density

Sodium metal anode 33

TABLE 2.4
A comparative assessment of nanostructured current collectors and 3D hosts to stabilize the Na metal anode

Na cell	Salt	Solvent	Approaches/Technologies/Methods to stabilize Na metal anode	Current density (mA/cm²)	Capacity (mA h/cm²)	Cycle life (h)	Pro's	Con's
Na-Na	1 M NaClO₄	EC/PC	Carbon felt [77]	5	2	100	Uniform Na nucleation hinders the volume change and effective suppression of Na dendrites	Large impedance (450 Ω in the 50th cycle and 550 Ω in the 120th cycle)
Na-Na	1 M NaClO₄	EC/DEC	Wood – carbon – Na [78]	0.5	0.25	250	High surface area anode, mechanically stable, uniform Na nucleation, and hinders the volume change	Cycle life could not be improved significantly
Na-Na	1 M NaPF₆	EC/PC	Na@CP - NCNT [79]	1	1	350	Uniform Na nucleation, more Na-phillic sites, stable electrochemical performance	The interphase lacks mechanical stability
Na/Al-Na/Al	1 M NaPF₆	DEGDME	Porous Al Foil [80]	0.5	0.25	1000	Large available surface for Na nucleation, homogeneous plating	An optimized pore structure is highly cumbersome and challenging
Na/Cu- Na/Cu	1 M NaPF₆	DEGDME	3D Cu nanowire [81]	0.5	0.5	400	Stable electrochemical deposition, long life cycle, and high efficiency	The deposition/dissolution overpotential increases with the cycle number
Na/Ni- Na/Ni	1 M NaPF₆	DEGDME	Porous 3D nickel foam [82]	1	0.5	220	Exceptional cyclic stability, reduced volume change	Cycle life could not be improved

Nanostructured current collector and host

(Continued)

TABLE 2.4 (CONTINUED)

Na cell	Salt	Solvent	Approaches/Technologies/Methods to stabilize Na metal anode	Current density (mA/cm^2)	Capacity (mA h/cm^2)	Cycle life (h)	Pro's	Con's
Na/C- Na/C	1 M NaCF$_3$SO$_3$	Diglyme	Carbon fiber [83]	1	1	200	High CE, dendrite-free Na anode	Cycle life could not be improved
Na/Cu- Na/Cu	1 M NaCF$_3$SO$_3$	Diglyme	Au on Cu [84]	2	1	300	Reduced nucleation overpotential, high CE even at a high current density	The thickness of the Au layer is not optimized well, which in turn consumes plenty of Na metal to form Au-Na alloy
Na-Na	1 M NaClO$_4$	EC/PC	RGO-Na [76]	0.25	0.0625	300	High-strength anode, increased hardness, and highly stable to corrosion	A large deposition/dissolution overpotential in carbonate-based electrolyte
	1 M NaPF$_6$	DEGDME	RGO-Na	1	1	600		
Na-Fe$_2$O$_3$@Ni	1 M NaPF$_6$	DEGDME	Fe$_2$O$_3$ @Ni [85]	5	1	500	Three-dimensional reticulation structure with approximately 300 μm meshes, which facilitates the diffusion of Na$^+$ and provides a considerable space for Na deposition	Cycle life could not be improved significantly

have seen rapid progress over the past two years. More fundamental advancements are highly anticipated with continued focus in this field.

In conclusion, employing just one strategy will not address all of the problems Na metal anodes are experiencing. The metallic Na anode will become a reality with the help of a multi-strategy approach with clear goals.

REFERENCES

1. Sun B, Xiong P, Maitra U, et al. Design Strategies to Enable the Efficient Use of Sodium Metal Anodes in High-Energy Batteries. *Adv Mater.* 2020;32:1903891.
2. Eng AYS, Soni CB, Lum Y, et al. Theory-Guided Experimental Design in Battery Materials Research. *Sci Adv.* 2022;8:eabm2422.
3. Sundaram PM, Soni CB, Sungjemmenla, et al. Reviving bipolar construction to design and develop high-energy sodium-ion batteries. *J Energy Storage.* 2023;63:107139.
4. Jishnu NS, Vineeth SK, Das A, et al. Electrospun PVdF and PVdF-co-HFP-Based Blend Polymer Electrolytes for Lithium Ion Batteries. In: Balakrishnan NTM, Prasanth R, editors. *Electrospinning for Advanced Energy Storage Applications.* 1st ed. Singapore: Springer Singapore; 2021. pp. 201–234.
5. Yan Z, Zhao L, Wang Y, et al. The Future for Room-Temperature Sodium–Sulfur Batteries: From Persisting Issues to Promising Solutions and Practical Applications. *Adv Funct Mater.* 2022;32:2205622.
6. Pozo-gonzalo C, Ortiz-vitoriano N. ScienceDirect Electrochemistry Recent Progress, Advances, and Future Prospects in Na–O_2 Batteries. *Curr Opin Electrochem.* 2022;36:101120.
7. Kim B-R, Jeong G, Kim A, et al. High Performance Na-$CuCl_2$ Rechargeable Battery toward Room Temperature ZEBRA-Type Battery. *Adv Energy Mater.* 2016;6:1600862.
8. Soni CB, Sungjemmenla, Vineeth SK, et al. Challenges in Regulating Interfacial-Chemistry of the Sodium-Metal Anode for Room-Temperature Sodium-Sulfur Batteries. *Energy Storage.* 2022;4:e264.
9. Wang Y, Zhou D, Palomares V, et al. Revitalising Sodium-Sulfur Batteries for Non-High-Temperature Operation: A Crucial Review. *Energy Environ Sci.* 2020;13:3848–3879.
10. Perkin FM. The Discovery of the Alkali Metals by Humphry Davy: The Bearing of the Discovery upon Industry. *Trans Faraday Soc.* 1908;3:205.
11. Sungjemmenla, Vineeth SK, Soni CB, et al. Understanding the Cathode–Electrolyte Interphase in Lithium-Ion Batteries. *Energy Technol.* 2022;10:2200421.
12. Sungjemmenla, Soni CB, Vineeth SK, et al. Exploration of the Unique Structural Chemistry of Sulfur Cathode for High-Energy Rechargeable Beyond-Li Batteries. *Adv Energy Sustain Res.* 2022;3:2100157.
13. Sungjemmenla, Soni CB, Vineeth SK, et al. Unveiling the Physiochemical Aspects of the Matrix in Improving Sulfur-Loading for Room-Temperature Sodium–Sulfur Batteries. *Mater Adv.* 2021;2:4165–4189.
14. Sungjemmenla, Soni CB, Kumar V. Recent Advances in Cathode Engineering to Enable Reversible Room-Temperature Aluminium–Sulfur Batteries. *Nanoscale Adv.* 2021;3:1569–1581.
15. Soni CB, Sungjemmenla, Vineeth SK, et al. Patterned Interlayer Enables a Highly Stable and Reversible Sodium Metal Anode for Sodium-Metal Batteries. *Sustain Energy Fuels.* 2023;7:1908–1915.
16. Chen X, Shen X, Hou T-Z, et al. Ion-Solvent Chemistry-Inspired Cation-Additive Strategy to Stabilize Electrolytes for Sodium-Metal Batteries. *Chem.* 2020;6:2242–2256.

17. Vineeth SK, Soni CB, Sun Y, et al. Implications of Na-Ion Solvation on Na Anode–Electrolyte Interphase. *Trends Chem.* 2022;4:48–59.
18. Jin Q, Lu H, Zhang Z, et al. Synergistic Manipulation of Na+ Flux and Surface-Preferred Effect Enabling High-Areal-Capacity and Dendrite-Free Sodium Metal Battery. *Adv Sci.* 2022;9:2103845.
19. Peled E, Golodnitsky D, Ardel G. Advanced Model for Solid Electrolyte Interphase Electrodes in Liquid and Polymer Electrolytes. *J Electrochem Soc.* 1997;144:L208–L210.
20. Goodenough JB, Kim Y. Challenges for Rechargeable Li Batteries. *Chem Mater.* 2010;22:587–603.
21. Shen X, Zhang R, Chen X, et al. The Failure of Solid Electrolyte Interphase on Li Metal Anode: Structural Uniformity or Mechanical Strength. *Adv Energy Mater.* 2020;10:1–8.
22. Medenbach L, Bender CL, Haas R, et al. Origins of Dendrite Formation in Sodium–Oxygen Batteries and Possible Countermeasures. *Energy Technol.* 2017;5:2265–2274.
23. Chazalviel JN. Electrochemical Aspects of the Generation of Ramified Metallic Electrodeposits. *Phys Rev A.* 1990;42:7355–7367.
24. Lin D, Liu Y, Cui Y. Reviving the Lithium Metal Anode for High-Energy Batteries. *Nat Nanotechnol.* 2017;12:194–206.
25. Wang H, Wang C, Matios E, et al. Critical Role of Ultrathin Graphene Films with Tunable Thickness in Enabling Highly Stable Sodium Metal Anodes. *Nano Lett.* 2017;17:6808–6815.
26. Kim YJ, Lee H, Noh H, et al. Enhancing the Cycling Stability of Sodium Metal Electrodes by Building an Inorganic-Organic Composite Protective Layer. *ACS Appl Mater Interfaces.* 2017;9:6000–6006.
27. Wang S, Jie Y, Sun Z, et al. An Implantable Artificial Protective Layer Enables Stable Sodium Metal Anodes. *ACS Appl Energy Mater.* 2020;3:8688–8694.
28. Lu K, Gao S, Li G, et al. Regulating Interfacial Na-Ion Flux via Artificial Layers with Fast Ionic Conductivity for Stable and High-Rate Na Metal Batteries. *ACS Mater Lett.* 2019;1:303–309.
29. Ding F, Xu W, Chen X, et al. Effects of Cesium Cations in Lithium Deposition via Self-Healing Electrostatic Shield Mechanism. *J Phys Chem C.* 2014;118:4043–4049.
30. Yui Y, Hayashi M, Nakamura J. In Situ Microscopic Observation of Sodium Deposition/Dissolution on Sodium Electrode. *Sci Rep.* 2016;6:1–8.
31. Cheng XB, Zhang R, Zhao CZ, et al. Toward Safe Lithium Metal Anode in Rechargeable Batteries: A Review. *Chem Rev.* 2017;117:10403–10473.
32. Hong YS, Li N, Chen H, et al. In Operando Observation of Chemical and Mechanical Stability of Li and Na Dendrites under Quasi-Zero Electrochemical Field. *Energy Storage Mater.* 2018;11:118–126.
33. Soni CB, Kumar V, Seh ZW. Guiding Uniform Sodium Deposition through Host Modification for Sodium Metal Batteries. *Batter Supercaps.* 2022;5:e202100207.
34. Zhang J, Wang W, Shi R, et al. Three-Dimensional Carbon Felt Host for Stable Sodium Metal Anode. *Carbon.* 2019;155:50–55.
35. Yabuuchi N, Matsuura Y, Ishikawa T, et al. Phosphorus Electrodes in Sodium Cells: Small Volume Expansion by Sodiation and the Surface-Stabilization Mechanism in Aprotic Solvent. *ChemElectroChem.* 2014;1:580–589.
36. Zhao C, Lu Y, Yue J, et al. Advanced Na Metal Anodes. *J Energy Chem.* 2018;27:1584–1596.
37. Fincher CD, Zhang Y, Pharr GM, et al. Elastic and Plastic Characteristics of Sodium Metal. *ACS Appl Energy Mater.* 2020;3:1759–1767.
38. Fan L, Li X. Recent Advances in Effective Protection of Sodium Metal Anode. *Nano Energy.* 2018;53:630–642.
39. Zheng X, Fu H, Hu C, et al. Toward a Stable Sodium Metal Anode in Carbonate Electrolyte: A Compact, Inorganic Alloy Interface. *J Phys Chem Lett.* 2019;10:707–714.

40 Stratford JM, Mayo M, Allan PK, et al. Investigating Sodium Storage Mechanisms in Tin Anodes: A Combined Pair Distribution Function Analysis, Density Functional Theory, and Solid-State NMR Approach. *J Am Chem Soc.* 2017;139:7273–7286.
41 Fang W, Jiang H, Zheng Y, et al. A Bilayer Interface Formed in High Concentration Electrolyte with SbF_3 Additive for Long-Cycle and High-Rate Sodium Metal Battery. *J Power Sources.* 2020;455:227956.
42 Zhu M, Li S, Li B, et al. Homogeneous Guiding Deposition of Sodium through Main Group II Metals toward Dendrite-Free Sodium Anodes. *Sci Adv.* 2019;5:1–9.
43 Zheng J, Zhao Q, Tang T, et al. Reversible Epitaxial Electrodeposition of Metals in Battery Anodes. *Science.* 2019;366:645–648.
44 Chen X, Shen X, Li B, et al. Ion–Solvent Complexes Promote Gas Evolution from Electrolytes on a Sodium Metal Anode. *Angew Chemie Int Ed.* 2018;57:734–737.
45 Rodriguez R, Loeffler KE, Nathan SS, et al. In Situ Optical Imaging of Sodium Electrodeposition: Effects of Fluoroethylene Carbonate. *ACS Energy Lett.* 2017;2:2051–2057.
46 Michalak B, Sommer H, Mannes D, et al. Gas Evolution in Operating Lithium-Ion Batteries Studied In Situ by Neutron Imaging. *Sci Rep.* 2015;5:1–9.
47 Jozwiuk A, Berkes BB, Weiß T, et al. The Critical Role of Lithium Nitrate in the Gas Evolution of Lithium-Sulfur Batteries. *Energy Environ Sci.* 2016;9:2603–2608.
48 Zhang B, Metzger M, Solchenbach S, et al. Role of 1,3-propane sultone and vinylene carbonate in solid electrolyte interface formation and gas generation. *J Phys Chem C.* 2015;119:11337–11348.
49 Vineeth SK, Tebyetekerwa M, Liu H, et al. Progress in the Development of Solid-State Electrolytes for Reversible Room-Temperature Sodium-Sulfur Batteries. *Mater Adv.* 2022;2:6415–6440.
50 Cao K, Ma Q, Tietz F, et al. A Robust, Highly Reversible, Mixed Conducting Sodium Metal Anode. *Sci Bull.* 2021;66:179–186.
51 Liu S, Bai M, Tang X, et al. Enabling High-Performance Sodium Metal Anode via a Presodiated Alloy-Induced Interphase. *Chem Eng J.* 2021;417:128997.
52 Gu Y, Wang WW, Li YJ, et al. Designable Ultra-Smooth Ultra-Thin Solid-Electrolyte Interphases of Three Alkali Metal Anodes. *Nat Commun.* 2018;9:1339.
53 Seh ZW, Sun J, Sun Y, et al. A Highly Reversible Room-Temperature Sodium Metal Anode. *ACS Cent Sci.* 2015;1:449–455.
54 Wang S, Chen Y, Jie Y, et al. Stable Sodium Metal Batteries via Manipulation of Electrolyte Solvation Structure. *Small Methods.* 2020;4:1900856.
55 Kumar V, Eng AYS, Wang Y, et al. An Artificial Metal-Alloy Interphase for High-Rate and Long-Life Sodium–Sulfur Batteries. *Energy Storage Mater.* 2020;29:1–8.
56 Shi Q, Zhong Y, Wu M, et al. High-Performance Sodium Metal Anodes Enabled by a Bifunctional Potassium Salt. *Angew Chemie.* 2018;130:9207–9210.
57 Zheng X, Gu Z, Liu X, et al. Environmental Science Extinguishant with Highly-Fluorinated Electrolytes. *Energy Environ Sci.* 2020;13:1788–1798.
58 Yi Q, Lu Y, Sun X, et al. Fluorinated Ether Based Electrolyte Enabling Sodium-Metal Batteries with Exceptional Cycling Stability. *ACS Appl Mater Interfaces.* 2019;11:46965–46972.
59 Zheng J, Chen S, Zhao W, et al. Extremely Stable Sodium Metal Batteries Enabled by Localized High-Concentration Electrolytes. *ACS Energy Lett.* 2018;3:315–321.
60 Luo J, Zhang Y, Matios E, et al. Stabilizing Sodium Metal Anodes with Surfactant-Based Electrolytes and Unraveling the Atomic Structure of Interfaces by Cryo-TEM. *Nano Lett.* 2022;22:1382–1390.
61 Dugas R, Ponrouch A, Gachot G, et al. Na Reactivity toward Carbonate-Based Electrolytes: The Effect of FEC as Additive. *J Electrochem Soc.* 2016;163:A2333–A2339.
62 Han M, Zhu C, Ma T, et al. In Situ Atomic Force Microscopy Study of Nano–Micro Sodium Deposition in Ester-Based Electrolytes. *Chem Commun.* 2018;54:2381–2384.

63 Zhao Y, Goncharova LV, Zhang Q, et al. Inorganic–Organic Coating via Molecular Layer Deposition Enables Long Life Sodium Metal Anode. *Nano Lett.* 2017;17:5653–5659.
64 Zhao Y, Goncharova LV, Lushington A, et al. Superior Stable and Long Life Sodium Metal Anodes Achieved by Atomic Layer Deposition. *Adv Mater.* 2017;29:1606663.
65 Luo W, Lin C, Zhao O, et al. Ultrathin Surface Coating Enables the Stable Sodium Metal Anode. *Adv Energy Mater.* 2017;7:1601526.
66 Tian H, Seh ZW, Yan K, et al. Theoretical Investigation of 2D Layered Materials as Protective Films for Lithium and Sodium Metal Anodes. *Adv Energy Mater.* 2017;7:1602528.
67 Jiang F, Li T, Ju P, et al. Nano-SiO_2 Coating Enabled Uniform Na Stripping/Plating for Dendrite-Free and Long-Life Sodium Metal Batteries. *Nanoscale Adv.* 2019;1:4989–4994.
68 Zhao Y, Liang J, Sun Q, et al. In Situ Formation of Highly Controllable and Stable Na_3PS_4 as a Protective Layer for Na Metal Anode. *J Mater Chem A.* 2019;7:4119–4125.
69 Zhu M, Wang G, Liu X, et al. Dendrite-Free Sodium Metal Anodes Enabled by a Sodium Benzenedithiolate-Rich Protection Layer. *Angew Chemie.* 2020;132:6658–6662.
70 Li P, Xu T, Ding P, et al. Highly Reversible Na and K Metal Anodes Enabled by Carbon Paper Protection. *Energy Storage Mater.* 2018;15:8–13.
71 Wei S, Choudhury S, Xu J, et al. Highly Stable Sodium Batteries Enabled by Functional Ionic Polymer Membranes. *Adv Mater.* 2017;29:1605512.
72 Choudhury S, Wei S, Ozhabes Y, et al. Designing Solid-Liquid Interphases for Sodium Batteries. *Nat Commun.* 2017;8:898.
73 Luo Z, Tao S, Tian Y, et al. Robust Artificial Interlayer for Columnar Sodium Metal Anode. *Nano Energy.* 2022;97:107203.
74 Liang J, Wu W, Xu L, et al. Highly Stable Na Metal Anode Enabled by a Multifunctional Hard Carbon Skeleton. *Carbon.* 2021;176:219–227.
75 Soni CB, Arya N, Sungjemmenla, et al. Microarchitectures of Carbon Nanotubes for Reversible Na Plating/Stripping Toward the Development of Room-Temperature Na–S Batteries. *Energy Technol.* 2022;10:2200742.
76 Wang A, Hu X, Tang H, et al. Processable and Moldable Sodium-Metal Anodes. *Angew Chemie.* 2017;129:12083–12088.
77 Chi S-S, Qi X-G, Hu Y-S, et al. 3D Flexible Carbon Felt Host for Highly Stable Sodium Metal Anodes. *Adv Energy Mater.* 2018;8:1702764.
78 Luo W, Zhang Y, Xu S, et al. Encapsulation of Metallic Na in an Electrically Conductive Host with Porous Channels as a Highly Stable Na Metal Anode. *Nano Lett.* 2017;17:3792–3797.
79 Zhao Y, Yang X, Kuo L-Y, et al. High Capacity, Dendrite-Free Growth, and Minimum Volume Change Na Metal Anode. *Small.* 2018;14:1703717.
80 Liu S, Tang S, Zhang X, et al. Porous Al Current Collector for Dendrite-Free Na Metal Anodes. *Nano Lett.* 2017;17:5862–5868.
81 Lu Y, Zhang Q, Han M, et al. Stable Na Plating/Stripping Electrochemistry Realized by a 3D Cu Current Collector with Thin Nanowires. *Chem Commun.* 2017;53:12910–12913.
82 Xu Y, Menon AS, Harks PPRML, et al. Honeycomb-Like Porous 3D Nickel Electrodeposition for Stable Li and Na Metal Anodes. *Energy Storage Mater.* 2018;12:69–78.
83 Zhang Q, Lu Y, Zhou M, et al. Achieving a Stable Na Metal Anode with a 3D Carbon Fibre Scaffold. *Inorg Chem Front.* 2018;5:864–869.
84 Tang S, Qiu Z, Wang XY, et al. A Room-Temperature Sodium Metal Anode Enabled by a Sodiophilic Layer. *Nano Energy.* 2018;48:101–106.
85 Jiang F, Li X, Wang J, et al. Long-Life and Efficient Sodium Metal Anodes Enabled by a Sodiophilic Matrix. *J Alloys Compd.* 2022;910:164762.

3 Sulfur cathode
Progress in the development of sulfur cathode

Sungjemmenla, S.K. Vineeth, and Vipin Kumar

3.1 INTRODUCTION TO SULFUR CATHODE

With the attenuation of demand in perceiving the high-capacity cathodes, sulfur as a conversion cathode has leapfrogged the traditional insertion-type cathodes (e.g., oxide-based cathodes) with a remarkable theoretical capacity of 1675 mAh g^{-1} based on the reaction of two-electron per sulfur atom [1–4]. Notably, sulfur has relatively positive reduction potential (compared to sodium metal), it is abundant in nature (16th most abundant element on the Earth's crust), is cost-effective (less than $2.0 kg^{-1}), and it is available as a byproduct of petroleum refineries [5]. While combined with suitable anode materials (e.g., Li, Na, or Mg) [6,7], it yields potentially high energy densities significantly higher than the expensive transition metal oxides [8,9]. Considered one of the highest-capacity cathodes among solid-state materials, sulfur provides the additional benefit of functioning as a thermal and electrical insulator with an extremely low electrochemical potential (0.407 V vs. SHE) [10]. Based on different polyatomic structures and oxidation states (−2, +2, +4, and +6), octasulfur (cyclo-S_8), an allotrope of sulfur, is regarded as the most stable at ambient conditions. When coupled with the sodium metal anode at room temperature, the battery can deliver a remarkably high specific energy density of about 1230 Wh kg^{-1} [11]. Thus, sulfur as a potential contender has been revitalized as a cathode in sodium-sulfur batteries [12]. Table 3.1 illustrates some of the properties of the elemental sulfur at ambient conditions.

3.1.1 Mechanistic Principle of the Sulfur Conversion Reaction

The elemental sulfur is highly insulating and, therefore, can't be directly used as the cathode material [13,14]. Generally, carbon matrix is used as a conductive filler to function elemental sulfur as the sulfur cathode [15–17]. The sulfur cathode, coupled with the alkali or alkali-earth metal, generally serves as a metal-sulfur battery [18]. A basic configuration of a metal sulfur battery, for instance, a sodium-sulfur battery, can be illustrated as depicted in Figure 3.1a. The choice of electrolyte plays a significant role in determining the stability of the cell [19], as the metal anode or sulfur cathode is not equally compatible with the organic solvent or electrolyte salt systems [20].

DOI: 10.1201/9781003388067-3

TABLE 3.1
Important properties of the elemental Sulfur

Sulfur	Atomic weight (a.m.u)	Electronegativity	Oxidation states	Conductivity (S cm^{-1}) @ 25°C	Potential (vs. SHE)	Theoretical capacity (mAh g^{-1})	Boiling point (°C)	Melting point (°C)	Density (g cm^{-3}) @ 20°C
	32.065	2.58 eV	−2, +2, +4, +6	5*10^{-30}	0.407	1675	444.6	112.8 (α-S) 118.7 (β-S)	2.07 (α-S) 1.96 (β-S)

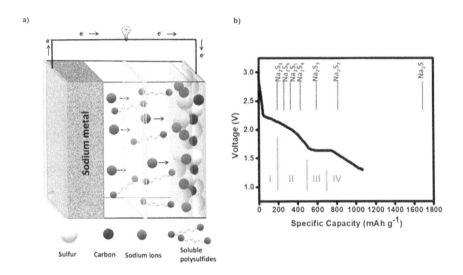

FIGURE 3.1 (a) A schematic illustration of a RT-Na/S battery technology. (b) Representation of voltage profile vs. theoretical-practical discharge capacities of a RT-Na/S battery.

During discharge, the overall electrochemical reaction occurring in a metal-sulfur battery can be expressed as follows:

Anode: $nM \rightarrow nM^{n+} + ne^-$
Cathode: $S_8 + nM^{n+} + ne^- \rightarrow M_xS$
Overall: $S_8 + nM \rightarrow M_xS$

During discharge reactions, the sulfur cathode undergoes a series of charge reactions (based on two-electron redox reactions). Simultaneously, metal stripping takes place at the anode. Four regions are of great interest based on the discharge curve and transition of sulfur to polysulfides. The electrochemical reactions of which are shown below (Figure 3.1b):

Scheme I: $S_8 + 2Na^+ + 2e^- \rightarrow Na_2S_8$ ~2.20 V vs. Na/Na$^+$

Region I: This region is considered the "high voltage plateau region" (at 2.20 V), which corresponds to the transition from S_8 to Na_2S_8 (solid-liquid transition region).

Scheme II: $Na_2S_8 + 2Na^+ + 2e^- \rightarrow 2Na_2S_4$ 2.20–1.65 V vs. Na/Na$^+$

Region II: This region is considered the "sloping region" (ranging from 2.20 to 1.65 V), which corresponds to the transition within the soluble polysulfides, i.e., Na_2S_8 to Na_2S_4 (liquid-liquid transition region). It is to be noted that this region is considered to be one of the most complex regions as the reactions are affected by the chemical equilibria among the soluble polysulfides coexisting in the electrolyte.

Scheme III: $Na_2S_4 + 2/3\ Na^+ + 2/3e^- \rightarrow 4/3\ Na_2S_3$ ~1.65 V vs. Na/Na^+

$Na_2S_4 + 2Na^+ + 2e^- \rightarrow 2Na_2S_2$

$Na_2S_4 + 6Na^+ + 6e^- \rightarrow 4Na_2S$

Region III: This region is considered the "low voltage plateau region" (at 1.65 V), which corresponds to the transition from soluble Na_2S_4 to insoluble Na_xS ($3 \leq x \leq 1$) (liquid-solid transition region). The complex steps involving the transition from dissolved polysulfides to insoluble ones determine the capacity and discharge voltages.

Scheme IV: $Na_2S_2 + 2Na^+ + 2e^- \rightarrow 2Na_2S$ 1.65–1.20 V vs. Na/Na^+

Region IV: This region is considered the "second sloping region" (ranging from 1.65 to 1.20 V), which corresponds to the transition within the insoluble polysulfides, i.e., Na_2S_2 to Na_2S (solid-solid transition region). This region suffers relatively sluggish kinetics due to the insulating nature of Na_2S and Na_2S, resulting in high polarization.

The multi-step complex steps in transforming sulfur to a series of sodium polysulfides result in sulfur reduction at the cathode during sodiation [21]. As a result, metal ions pass through the electrolyte to form sodium polysulfides and electrons via the external circuit [22]. Please note that the polysulfide formation is essential to harness the full potential of the sulfur cathode; however, its dissolution into the electrolyte is undesirable as it leads to several undesirable phenomena, such as the shuttle effect, capacity fade, and cell failure. In addition, it is the conversion of the long-chain polysulfide ($Na_2S_8 - Na_2S_6 - Na_2S_4$) to the short-chain polysulfide ($Na_2S_3 - Na_2S_2 - Na_2S$) which causes most of the irreversibility in the battery and eventually leads to dropping in the capacity of the cell. Efforts are made to rationally design a suitable sulfur cathode to minimize the polysulfide dissolution or polysulfide shuttling without affecting the cell's performance.

3.2 CHALLENGES IN DEVELOPING THE STABLE SULFUR CATHODES

Several challenges from both fundamental and redox kinetic reactions are encountered that obstruct the pathway to propel the growth toward a viable sulfur-based battery system [23]. In addition to their similar working mechanism, metal-sulfur battery chemistries are addressed with comparable issues.

3.2.1 INSULATING NATURE OF SULFUR AND ITS DISCHARGE SPECIES

Despite the high sustainability and non-toxicity of sulfur, one of the main bottlenecks that remain in this cathode system is the insulating nature of sulfur with very low conductivity, i.e., 5×10^{-30} S cm^{-1} at 25°C and its insoluble reduction products (Na_2S/Na_2S_2) [24–26]. As a result, ion/electron movement is impeded, low sulfur utilization is achieved, and a low amount of active materials can be loaded onto the electrode materials, further compromising the energy density of the battery chemistry [27,28].

3.2.2 Polysulfide Shuttling

Cyclo-S_8 undergoes a series of reactions involving a chain of intermediate redox species during the discharge process. The involved multi-step reduction of sulfur species leads to the generation of high-order polysulfides Na_2S_y (4 ≤ y ≤ 8) and low-order polysulfides Na_2S_y(4 ≤ y ≤ 1) [29,30]. The solubility of intermediate polysulfides differs with different liquid electrolyte systems [31]. The long-chain polysulfides are soluble in certain organic-based electrolytes and can diffuse via a separator, allowing them to shuttle quickly between the cathode and the anode. When the soluble polysulfides migrate toward the anode compartment, the species react with the metal (at the negative electrode) to reduce to insoluble low-order metal polysulfides (Na_2S_2/Na_2S). The cascade of polysulfides diffuses back to the cathode and gets re-oxidized when the polarity is reversed. This movement of sulfur species between the anode and the cathode is called polysulfide shuttling [32]. The dissolution and re-deposition of sulfur species involve complex steps and phase-change behavior, which leads to kinetically slow electrochemistry and sharp attenuation of capacity values [33–35].

3.2.3 Volume Expansion

Generally, sulfur undergoes typical conversion reactions that entail pulverization of cathodes during charge/discharge cycles. Due to the repeated dissolution and deposition of sulfur and its discharge products, significant volume changes occur, which involve the deterioration of the morphological and structural integrity of the matrix composite [36]. Besides disrupting the sulfur cathode's architecture due to high volumetric fluctuations, the cell experiences degradation and instability in its cyclic performance [37].

3.2.4 Loss of Sulfur Species

During shuttling phenomena, the formation of low-order metal polysulfides at the anode becomes concentrated and hence, gets deposited, resulting in low uptake of active material. Furthermore, with prolonged cycling, precipitation of insoluble species occurs, resulting in the loss of active material [38–40].

3.2.5 Self-Discharge

Another technical challenge that impedes the practicality of the system is high self-discharge. Due to the high solubility of the reaction products of metal polysulfides, self-discharge becomes more prominent in such technologies [41,42]. The ease in the dissolution of active materials in the electrolyte leads to concentration differences in the system, allowing self-discharge to dominate when the polysulfides migrate toward the anode. The self-discharge can further be boosted on high elevated temperatures that can eventually result in limited cycling life [43].

3.2.6 Low Coulombic efficiency

As a conversion cathode, sulfur demonstrates a higher potential hysteresis between the discharge and charge processes that can result in sluggish electrochemical reaction kinetics. Due to lowered electroactivity, irreversible losses, and a shuttle mechanism, the battery system suffers from increased polarization. As a result, the properties of cells are impeded with rapid capacity fading rate and impedance during cycling accompanied by low values of Coulombic efficiencies [44].

3.3 PROGRESS IN DEVELOPING SULFUR CATHODES

To date, good progress has been made toward realizing a stable sulfur cathode in RT Na-S battery systems, as illustrated in Figure 3.2. Researchers have focused on a different combination of matrices and strategies to minimize the dissolution of polysulfides and enhance the kinetics of the redox reaction. The following approaches have been put forward to achieve a stable sulfur cathode.

3.3.1 Physically adsorbed S-host

The inherent insulating nature of sulfur and weak adsorption of the polysulfides with carbon matrices, a conductive host that possesses low electronic resistance (1–10 Ω), surface specific area (100–1000 m^2 g^{-1}), and high thermal and chemical stability is explored in the recent past [17,45,46]. Given the challenges mentioned above, cotton textile was utilized as a precursor to fabricate carbon fiber cloth (CFC), and it is used as a host for the sulfur cathode in RT Na-S battery. CFC offers high electronic conductivity and abundant surface sites to entrap the polysulfides. In addition to that, CFC

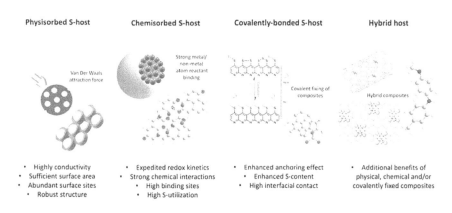

FIGURE 3.2 Overview of various hosts for S-cathode for RT-Na/S battery technology.

does not require polymeric binders to hold sulfur particles. A binder-free composite of CFC and sulfur was synthesized with different sulfur-loadings of 1 mg cm^{-2} (CFC/S-1), 2 mg cm^{-2} (CFC/S-2), and 3 mg cm^{-2} (CFC/S-3) [47]. Besides that, CFC provides pathways for better electrolyte absorbability, and the inter-fiber region serves as the buffer space for volumetric fluctuations. In contrast to CFC/S-3 and CFC/S-3, CFC/S with an optimum loading depicted the best performance with 1st discharge capacity of about 486 mAh g^{-1} and retaining a specific capacity of about 148 mAh g^{-1} upon 300 cycles. A surface-modified carbon felt was also used as a physical and electrical support for sulfur particles [48]. RT Na-S cells could survive only for 70 cycles, which could be due to the gradual dissolution of the polysulfide into the electrolyte medium.

The design of nanostructured host materials requires a complex synthesis procedure, in addition to the high cost and safety aspect [49,50]. To minimize the cost, processed carbon-based sulfur hosts are considered. Owing to a higher number of defects, the reversible capacity of processed carbon/commercial sulfur cathode could be enhanced significantly. The increased capacity could be credited to the increased number of adsorption sites (due to vacancy defects), which facilitate binding with sulfur. As shown in Figure 3.3a, the adsorption energy (E_{ad}) reaches the highest (–0.628 eV) when the size of the defects is 2, as compared to lower E_{ad} without carbon defects. Note that defect sizes 1 to 3 offer the highest binding effect toward the sodium polysulfides. This enables an enhanced binding which allows the facilitation of gaining more electrons by sulfur during the cycling process. Figure 3.3b shows the deformation charge density of a sulfur molecule adsorbed on various defect sites of a processed carbon surface [51].

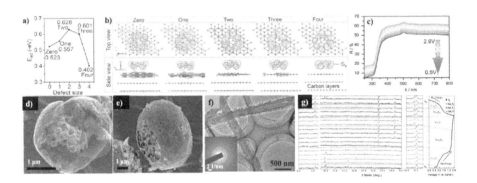

FIGURE 3.3 (a) Adsorption energy (Ead) of S8 vs. carbon defect sites (0 to 4). (b) Deformation charge density of carbon surface with different defect sites at an S8 adsorption site, Adapted and reproduced with permission from Ref. [52]. Copyright © 2020 Elsevier. (c) In situ UV/vis spectra of ACC/S cathode with 40 wt.% sulfur, Adapted and reproduced with permission from Ref. [53]. Copyright © 2020 John Wiley and Sons. SEM images of (d) PCMs, and (e) PCMs/S, Adapted and reproduced with permission from Ref. [54]. Copyright © 2018 American Chemical Society. (f) TEM and Selected area electron diffraction (SAED) patterns of CHNBs@PCNFs, Adapted and reproduced with permission from Ref. [55]. Copyright © 2018 Elsevier. (g) In situ operando XRD patterns of iMCHS cell based on 1st charge/discharge curve. Adapted and reproduced with permission from Ref. [56]. Copyright © 2016 American Chemical Society.

In addition to the processed carbon as a physical host for the sulfur cathode, microporous carbons are considered for their small pore sizes and high surface area [56–58]. These are incredibly effective hosts for various metal-sulfur batteries, including RT Na-S batteries. Since the pore volume of the microporous carbon is much smaller than the volume of the S_8 molecule (3.1 nm^3), the linear sulfur chain molecules fill the pores. Microporous carbon (<2 nm) with homogeneous pores are containers for accommodating and immobilizing sulfur and polysulfides. This host, however, is intrigued by the limited amount of sulfur loading inside the matrix. Owing to the microporous carbon host, the sulfur cathode could deliver a high specific capacity of 300 mAh g^{-1} even after 1500 cycles with a high Coulombic efficiency of about 98% [59].

A polyhedron-shaped microporous carbon that offers multi-facet sulfur adsorption is used as the sulfur host [60]. The unique morphology of a rhombic dodecahedron allowed for retaining most of the polysulfides from shuttling in the RT Na-S battery during cycling. It is a fact that the formation of polysulfide is accountable for obtaining a higher capacity, but polysulfide causes severe capacity fade; examining their deposition on the cathode surface could able to extend the cycle life of RT Na-S battery systems. More recently, Wang et al. [61] examined short-chain polysulfides, particularly, Na$_2$S deposition behavior, where they employed microporous-based carbon spheres as the host for the sulfur cathode. Through manipulating the insoluble polysulfides, a self-supporting cathode system could be achieved with uniform distribution of sulfur and facilitated electron transfer. The sulfur cathode could retain a capacity of about 63.4% with a decay of 0.07% per cycle. The postmortem analysis unveiled that the as-developed cathode facilitates a uniform and smooth accumulation of polysulfides on the surface, resulting in maximum sulfur utilization.

Generally, microporous carbon hosts are activated to ensure the opening and availability of the pores [62,63]. A diluted acid (1–2 M HCl) is commonly used to activate the pores chemically. An activation-free approach for synthesizing microporous and ultra-microporous carbon (d < 0.7 nm) has also been developed recently [64]. Though the distributed pore size of ultra-microporous carbon allowed good affinity among the sulfur species with the carbon scaffold, effectively suppressing possible side reactions of sodium polysulfides, the system becomes less conductive and eventually resulting in lower utilization of active material at higher sulfur loading (>3 mg cm^{-2}). Therefore, activation of the microporous carbon is essential to perform the process; however, the activation time, sometimes referred to as the etching time, is not yet standardized. Surprisingly, Guo and co-workers worked on an activated ultramicroporous coffee carbon (ACC) successfully incorporating nearly 40 wt.% of sulfur inside the slit ultramicropore carbon backbone [52]. Based on first principle calculations, they could correlate the slit micropore's space confinement with the dissolution of polysulfides. Additionally, operando UV-vis spectroscopy was conducted to demonstrate Na$_2$S as the final discharge product with effective alleviation of polysulfides at the end of the discharge reaction (Figure 3.3c).

Despite the excellent encapsulation of polysulfide species inside the microporous matrices, the loss of discharge moieties during cycling is inevitable due to the direct contact of embedded polysulfides with the electrolyte. To overcome this, Zhang et al. [53] focused on a porous yet sealed structure of "double-carbon-shell passion fruit-structured porous carbon microspheres (PCMs)." The PCMs were composed of an outer shell of micro-sized carbon and an inner shell of hollow nanobeads, as shown in Figure 3.3d. After sulfur infiltration, the morphology of the PCMs could be well-retained, which explains the crucial role of double-shell carbon structure (Figure 3.3e). Owing to the unique structure of PCMs, the cell could deliver an initial discharge capacity of 1100 mAh g^{-1} at a current density of 100 mA g^{-1} and retains a capacity of about 290 mAh g^{-1} after 350 cycles. Note that the soluble polysulfides were well-protected inside the spheres, allowing minimum contact with the electrolyte, and the voids provided sufficient space for volumetric fluctuations. Because of the unique characteristics of porous carbon backbone, hollow carbon nanobubbles on porous carbon nanofibers (CHNBs@PCNFs) were demonstrated [54]. The homogenous morphology of carbon nanobubbles reveals the hollow structure of carbon nanobubbles and thin shells of ~2 nm thickness, where the SAED patterns illustrate the amorphous nature of the composite material (Figure 3.3f). Successful infiltration of sulfur was accompanied by the long fibrous structure of CHNBs, with no obvious aggregated sulfur particles on the surface, resulting in an enormous percentage of sulfur (71.2%) inside the void spaces of CHNBs.

Interconnected and interlaced mesoporous carbon hollow nanospheres (iMCHS) with good structural integrity and high tap density are considered promising as the host for the sulfur cathode [55]. Based on the in situ synchrotron XRD studies (Figure 3.3g), reversibility was observed during the reactions between elemental sulfur and Na_2S_4. Meanwhile, the irreversibility of solid Na_2S can be observed based on the diffraction peak at (220) corresponding to Na_2S, which tells us that the further reduction into polysulfides was obstructed. The hollow carbon backbone allowed sufficient spaces to accommodate large amounts of sulfur, resulting in a mass loading of 3.2–4.1 mg cm^{-2}.

3.3.2 Chemisorbed S-host

The physical confinement of molecular sulfur species (i.e., S_8 molecules) into the non-polar carbon hosts is held together with a weak physical interaction, which becomes futile at higher sulfur loading [65,66]. The physically confined sulfur cathodes are often realized with low sulfur loadings (1–3 mg cm^{-2}) with light binding energies to suppress and immobilize sulfur and its intermediate products. Though a physically confined sulfur cathode can support a few tens of cycles at higher sulfur loading, polysulfide shuttling becomes more severe. The weak van der Waals force between the carbonaceous-based compounds and polysulfides has failed to restrain the poor reversibility, sluggish kinetics, and fast attenuation rate [67,68]. Various strategies have been developed to date to incorporate the higher amount of sulfur with minimized shuttle effect via chemical binding effect, which are as follows:

3.3.2.1 Heteroatom doping

Minimizing shutting and containing more active material could be realized by introducing polar heteroatoms into the carbon host [69,70]. The composite promotes good chemical interactions between carbon matrix and sulfur, enhancing charge transfer kinetics [71]. Heteroatoms with electronegativity close to or lower than carbon, for instance, nitrogen [72–74], phosphorus [75,76], and boron [77,78] are found to accelerate polysulfide conversion kinetics. However, using phosphorus or boron dopants is limited in studies because of their complex synthesis procedures. Intrigued by the doping of heteroatoms, Qiang et al. [79] reported a nanoporous carbon with high doping of N, S (~40 atom %) for RT Na-S batteries. As a result of the faster conversion kinetics and entrapment of polysulfides in nonporous carbon, the sulfur cathode could be cycled for about 8000 cycles at 4.6 mA g^{-1}.

In the quest to explore higher sulfur-accommodated composites, another group proposed using nitrogen as the dopant in graphene nanosheets/ sulfur nanocomposites (NGNS/S) [80]. The as-prepared composite showed a hierarchical morphology with a 3D viole-like structure with different S-content of 25, 45, 65, and 86 wt.%. Compared to the higher loaded sulfur content, NGNS/S-H25, with the lowest loading, ran for 300 cycles, delivering a discharge capacity of 48 mAh g^{-1} at 0.1 C, which exhibited the best performance. Furthermore, they suggested increasing the content of dopants with modification of structural properties to achieve the best performance with the high sulfur content.

Despite the excellent adsorption capability of non-metal dopants with strong polar-polar interactions toward sodium polysulfides, their electrochemical performance does not meet the practical applications. Besides, their interactions with the polysulfides are weaker than the metal components.

3.3.2.2 Single metal atoms

Furthermore, to enhance the chemical polar affinity toward sodium polysulfides, single metal atoms have been introduced to tackle the shuttle phenomena effect [81,82]. Due to their high polar-polar interactions, single metal atoms or clusters are known to reach the utmost atomic utilization. They possess high conductivity and expedite the catalysis of polysulfide conversion from long-chain polysulfides to Na_2S [83]. Due to their high physicochemical adsorption, studies have shown that single metal nanoclusters are known to narrow down sodium polysulfides' decomposition energy barrier to accelerate the battery chemistry's electrokinetic [84]. Thus, ensuring an effective pathway to enhance the electrochemical properties of RT Na-S battery.

Hence, in continuation with the developments in using single-atom metals, various transition-metal atoms have also been utilized in RT Na-S batteries. Transition

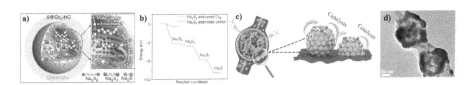

FIGURE 3.4 (a) A scheme presenting the redox reaction mechanism of S@Con-HC. (b) Energy profiles of Na_2S_4 adsorbed on carbon/Co and carbon support, Adapted and reproduced with permission from Ref. [88]. Copyright © 2018 Springer Nature. (c) Schematic illustration of catalytic effect of Ni-NCFs@S. (d) TEM image of Ni-NCFs@S, Adapted and reproduced with permission from Ref. [89]. Copyright © 2019 John Wiley and Sons.

elements such as iron (Fe), copper (Cu), and nickel (Ni) wreathed onto hollow carbon nanospheres (S@M-HC) were reported to create a chemical couple between sulfur and the metal atoms attenuating the dissolution of sodium polysulfides. Comparatively, the iron metal nanocluster composite delivered unprecedented results with effective entrapment of sulfur species and improved the catalytic activity to enhance the transformation of soluble to insoluble polysulfides [85]. For instance, to efficiently harness the attributes of single metal atoms, atomic cobalt (Co) as an electrocatalyst was studied for its efficiency. The investigators engineered the electrocatalyst by decorating Co in hollow carbon nanospheres (S@Co_n-HC) and employed it for RT Na-S battery [86]. The constructed sulfur cathode was realized with a thick layer of up to 5 mg cm^{-2} with successful sulfur encapsulation and its reduction species in the hollow carbon frameworks. This was possible with effective chemisorption by atomic cobalt single atom, as shown in Figure 3.4a. S@Co_n-HC as cathode retained an excellent reversible capacity of 508 mAh g^{-1} at 100 mAh g^{-1}, despite 600 cycles. Effectively high adsorption energy with a relatively high negative energy of -10.67 eV was observed in Figure 3.4b, indicating a shuttle-free phenomenon that followed a fast kinetic mechanism from intermediates to short-chain polysulfides.

The further enhancement of loading of active materials was visualized with a concept of chemical binding effect in a 3D matrix [12]. 3D matrices provide a highly accessible area to support the components, thereby accommodating many active materials. For instance, a hybrid structure of a 3D network composed of nitrogen-doped nickel hollow spheres CNFs (Ni-NCFs) effectively adsorb short-chain polysulfides, as schematically represented in Figure 3.4c. This can be well-affirmed by TEM analysis (Figure 3.4d), which depicts the successful accommodation of sulfur along the inner Ni hollow sphere represented by the darker color. A minimal decay rate of 0.17% per cycle was realized in addition to excellent rate capability at various C rates from 0.2 to 5 C [87]. In another study, an iron metal atom loaded onto a nitrogen-doped carbon nanosphere was employed as a matrix [88]. A fraction of iron (0.14 atomic wt.%) on the carbon scaffold effectively provided a high electrocatalytic activity to enhance the kinetic reaction.

In the pursuit of engineering the matrix, Ma et al. [89] reported a stable framework of graphene aerogel with carbon-wrapped cobalt nanoparticles (S@Co/C/rGO) with an S-content of about 37.5 wt.%. The dual function of graphene aerogel and cobalt clusters could provide a stable, interconnected architecture for fast electron/ion transfer and cobalt atoms and allow effective catalysis for enhanced kinetics during cycling, respectively. S@Co/C/rGO delivered remarkable reversible capacities at various C-rates; even at an exceptional rate of 5 C, the cathode still exhibited a discharge capacity after 1000 cycles with an attenuation rate of 0.01% per cycle. Similarly, the incorporation of gold and nitrogen-doped carbon microspheres maintained a prolonged cycle life of 2000 with a reversible capacity of 369 mAh g^{-1} at 10 A g^{-1}. Besides enhanced conversion kinetic reaction, the gold nanoclusters enabled strong chemical adsorption toward sulfur and its moieties, resulting in a high S-content of 56.5 wt.% [90].

Another approach to altering the architecture biologically was formulated by Du and co-workers, where they designed "a 3D branch leaf biometric design" [91]. The conductive branch leaf design was fabricated from cobalt nanoclusters on CNFs to enable the free migration of electrons and electrolyte uptake. The unique Co-S-Na interaction bonds could ensure fast conversion reduction reactions, delivering an impressive initial reversible capacity of 1201 mAh g^{-1} at 0.1 C. In 2020, Ye et al. [28] studied the electron states of nickel centers in a 2D MOF framework, which enabled the tuning of interactions between the polysulfides and the MOF matrix. A cathode with an S-loading of about 2.6 mg was prepared with an ultra-long cycle test of 1000 cycles retaining remarkable specific capacities at different C-rates. Lower voltage hysteresis was observed between the initial positive and negative peaks, which implies faster electrochemical reaction kinetics with a lower polarization effect.

Besides, the incorporation of metal atoms to complex polar metal phosphides provides strong electrocatalyst effects, which have been explored up to some extent. For instance, based on the structure of hetero-seed MOF, "a multi-region Janus featured CoP-Co structure" was introduced through a sequential carbonization-oxidation-perspiration process [92]. While the heterostructured composite with CoP-Co enhanced in optimizing the redox mechanism, the structured network was provided via nitrogen-doped carbon nanotube hollow nanocages (Figure 3.5a and b). Considering the dual synergistic effects, S@CoP-Co/NCNHC cathode delivered a prolonged cycle of 700 with a reversible discharge capacity of 448 mAh g^{-1} at 1 A g^{-1}. Nonetheless, the practical aspect of a commercial battery could not be met due to a realized low active mass loading of 1.7 mg cm^{-2}.

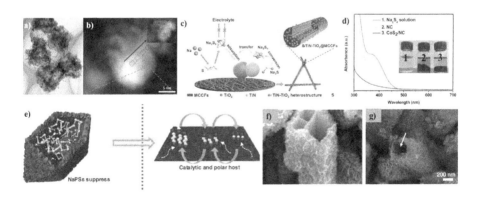

FIGURE 3.5 (a) STEM and (b) Janus featured area observed by HAADF-STEM of the S@CoP-Co/NCNHC composite material, Adapted and reproduced with permission from Ref. [93]. Copyright © 2020 American Chemical Society. (c) Schematic illustration of effective conversion reaction of polysulfides using TiN-TiO2, Adapted and reproduced with permission from Ref. [94]. Copyright © 2021 American Chemical Society. (d) UV-vis spectra with visible adsorption test of Na$_2$S$_4$ solution before and after exposure to NC and CoS2/NC, Adapted and reproduced with permission from Ref. [95]. Copyright © 2021 American Chemical Society. (e) Illustration of catalytic effect with effective polysulfide suppression in CoS2/C@S polar host, FESEM micrographs of (f) CoS2/C and (g) CoS2/C/S. Adapted and reproduced with permission from Ref. [96]. Copyright © 2020 Springer Nature.

3.3.2.3 Metal oxides

In addition to the use of non-metal and metal dopants, various other metal-based composites, such as metal oxides, have also been engineered in RT Na-S battery technologies [95,96]. Polar metal-based compounds can be engineered in metal-sulfur batteries as electrocatalysts to enhance the conductivity of sulfur and chemically immobilize the polysulfides [97]. An important aspect is a strong polarity and high conductivity, which can promote redox reactions in the active sites of the metal-based compounds [98,99]. The electrode should have a route to augment the conductivity of the system without compromising the performance of the battery. Despite high polar active sites, generally, metal oxides are non-conductive. Therefore, the rational design of different structures in combination with metal oxides provides state-of-the-art electrodes.

For instance, Ma et al. [100] reported a flexible freestanding cathode of a porous carbon matrix with a judicious combination of barium titanium trioxide ($BaTiO_3$) and amorphous titanium dioxide (TiO2) deposition on its surface. As one of the issues is to increase the polarity, a cost-effective and common $BaTiO_3$ with ferroelectric effect played an essential role in forming spontaneous polarization. Additionally, the synergistic structure of $BaTiO_3$ and TiO_2 on the matrix resulted in high sulfur encapsulation with high loading of up to ~3.5 mg cm^{-2}. Likewise, Ye et al. [101] reported a novel sulfur cathode of distinct heterostructures of TiN-TiO_2 to build on multichannel CNFs with an immediate electrocatalysis effect. While TiN provided an enhanced electrocatalytic activity, TiO_2 on the composite provided an additional benefit of strong chemisorption toward polysulfides, as illustrated in Figure 3.5c. Despite the unique characteristic structure of the matrix with expedited adsorption energy and electronic conductivity, the electrode was, however, plagued with a low amount of sulfur loading (1.08 mg cm^{-2}).

3.3.2.4 Metal sulfides

To date, there have been progress toward utilizing a metal sulfide-based system owing to its high catalytic effect and additional sulfiphilic property [102,103]. Exploiting the advantages of the metal sulfides in a battery system, Ye and co-workers introduced a 2D chain-like amorphous MoS_3 with an exceptionally high mass loading (7.1 mg cm^{-2}). MoS_3 as a cathode could maintain a stable long-term cycling performance of 1000 cycles at a current density of 0.45 A g^{-1} with a retained reversible capacity of ~180 mAh g^{-1} [104]. Another similar contribution was set forth by Meyerson et al. [105], where they introduced an amorphous molybdenum sulfide (MoS_x) with varying sulfur ratios. In contrast to $MoS_{3.5}$, $MoS_{5.6}$ yielded a better capacity of >300 mAh g^{-1} at a current density of 0.2 A g^{-1} over 100 cycles.

In the preceding year, Yan et al. [106] studied nickel sulfide nanocrystals with a high electrocatalysis effect decorated on porous CNTs doped with nitrogen with an areal loading of 2.5 mg cm^{-2}. Due to the physical and chemical entrapment of the sulfur, the composite sulfur cathode could successfully restrain the volumetric variations and mitigate polysulfide shuttling. In addition, a higher adsorption affinity of Na_2S toward NiS_2 with the binding energy of ~2.4 eV compared to N-doped CNTs (0.57 eV) demonstrates the fast conversion from long-chain polysulfides to short-chain Na_2S_2/Na_2S. Conversely, another group investigated the multi-sulfiphilic properties of other metals-sulfur systems. A core-shell architecture of bimetallic sulfides of ZnS and CoS_2 with nitrogen-doped carbon and sulfur encapsulated inside the matrix was fabricated using Zn/Co MOF precursors [107]. Apart from the intrinsic benefit of strong multi-sulfiphilic sites of metal sulfides, the core-shell architecture served as a reactor to store sulfur with a large amount of sulfur (~57 wt.%). With their effective catalysis and multiple active sites, the charging energy barrier of insoluble polysulfides is significantly reduced, thus concomitantly restricting the soluble polysulfides.

In 2021, Xiao and co-workers utilized the catalytic effect of CoS_2, where the short-chain sulfur species could be well-confined within the carbon matrix, constituting large pores greater than 0.5 nm [93]. A mesoporous structured CoS_2 based on MOF precursor/ nitrogen-doped carbon matrix (CoS_2/NC) was considered the host. As a catalyst, CoS_2 could weaken the S-S bond in the cyclo-S_8 structure, which was based on DFT calculations. UV-visible adsorption test of Na_2S_4 was conducted to evaluate the overwhelming polysulfide adsorption capability of CoS_2/NC compared to poor adsorption of NC without CoS_2 (Figure 3.5d). This enabled skipping the formation of soluble sodium polysulfides which facilitated the formation of insoluble polysulfides. Another unique metal chalcogenide architecture of "hollow, polar and catalytic bipyramid prisms of cobalt sulfide" was introduced by Aslam and co-workers [94]. Considered an effective S-host with an exorbitant catalytic effect, CoS_2 could successfully accelerate the conversion kinetics with an induced high conductivity. Besides, as schematically represented in Figure 3.5e, the polar catalytic CoS_2/C@S composite acts like a filter and provides strong interactions with polar sodium polysulfides to suppress the dissolution of polysulfides efficiently. Based on FESEM images (Figure 3.5f and g), the bipyramidal hollow architectures with wide open spaces (376 nm) of CoS_2/C and CoS_2/C/S can be observed and depicts the presence of sulfur inside the hollow matrix. The cell showed minimal decay in the capacity (0.0126% per cycle) even after 800 cycles.

3.3.2.5 Polysulfide catholyte

The reported contributions, nevertheless, still face critical challenges, including inhomogeneous distribution of sulfur and groundbreaking sodium storage capacities. Accordingly, researchers have focused on a unique approach by configuring a dissolved metal-polysulfide as the cathode. As a proof-of-concept, Rauh and co-workers proposed for the first time a lithium-dissolved polysulfide cathode be assembled in batteries for electrochemical tests, which dates back to 1979 [108]. However, their novel effort came to naught when the system experienced severe decay in capacity. Not long ago, Yu et al. [109] developed the unique structure of Na_2S as a cathode for the RT Na-S battery system in pursuit of an ideal host for the S-cathode. To overcome the inertness of Na_2S, the cathode particles were wrapped with multi-walled carbon nanotubes (MWCNT) fabrics, and their analysis revealed the formation of a cascade of intermediate polysulfides during cycling. Nevertheless, the system was impeded by limited charge/discharge cycles and the high cost of MWCNTs. Thus, the following year, they proposed a cost-efficient activated carbon nanofiber (AC-CNF) [110]. AC-CNF, with a self-weaving property, could form a stable framework for the

Na$_2$S cathode and provide a pathway for an efficient system. They could yield a high energy density of 790 Wh kg^{-1} based on active mass.

Later, Wang et al. [111] employed a conductive hierarchical carbon matrix with a spongy texture where Na$_2$S was incorporated into its framework, resulting in a structure similar to captivating frogspawn coral morphology. According to their studies, a shortened diffusion ion pathway was achieved together with facilitated migration of electrons in the cell. A new approach using the carbothermal reduction method of Na$_2$SO$_4$ at various temperatures was employed to result in a Na$_2$S/C cathode for an RT Na-S battery [112]. Na$_2$S/C delivered a high capacity utilization of 740 mAh g^{-1} with a stable number of 36 cycles, one of the highest to date based on Na$_2$S cells. Intrigued by the scope of this, Kumar et al. [113] experimented with a different configuration of using long-chain polysulfides as the catholyte instead of short-chain polysulfides. They have proposed Na$_2$S$_6$ catholyte as the active material embedded in a carbon framework of MnO$_2$ nanoarrays on carbon cloth. Based on an average voltage of 1.82 V, the as-assembled cell delivered a remarkable energy density of 946, 855, and 728 after the 1st, 50th, and 500th cycles, respectively.

3.3.3 Covalently bonded S-host

Due to the weak van der Waals adsorption of carbonaceous-based composites with sulfur, covalently bonded materials have certain functional groups, which increases the application of functional media to chemisorb polysulfides effectively [114,115]. Besides, during the physical confinement of carbon-sulfur composites, aggregate particles form, thus underutilizing the carbon scaffolds. Through covalent fixing, an enhanced anchoring effect is developed among the intermediate products without compromising the specific capacity of the battery [21]. Likewise, a controlled chain length of a covalently fixed composite of benzenedisulfonic acid with long-chain sulfur-carbon complex (SA-BDSA) was developed where –SO$_3$H and SO$_4^-$ were relied upon as the source for sulfur [116]. According to their studies, an insoluble species of RS$_2$O$_2^{2-}$ can be formed on the surface of the SA-BDSA composite, which can behave as a mediator to facilitate the interaction of soluble long-chain sodium polysulfides to insoluble surface-bound polysulfides. Consequently, forming O-S/C-S bridge bonds allows high interfacial contact and conductivity among the sulfur-carbon composites. At 2500 mA g^{-1}, SC-BDCA maintained a low fading rate of 0.035% per cycle for over 1000 cycles.

Considering the effect of covalent sulfur-carbon composite, Yan et al. [117] formulated an in situ characterization using a wet-solvothermal strategy between CS$_2$ and red phosphorus. The interior sulfur was immobilized and activated (under 0.5 V) by specifically designed covalent interactions. As a performance enhancer, sufficient accessibility was provided for sodium ions to migrate to the positive electrode due to the enlarged spacing of carbon to 0.4 nm after the initial cycle (Figure 3.6a). The covalent C/S cathode exhibited an outstanding electrochemical performance even after 950 cycles at 1.6 C. With strong adsorption capability toward polysulfides, the cathode delivered the best specific capacity of 700 mAh g^{-1} at a high rate of 8.1 C. Another strategy was proposed by Chen et al. [118] where they reported a sulfur-carbon composite to form a thioether-bonded functionalized carbon (SC). The thioether bond was constructed using a sulfur-rich precursor of thiophene monomer and a crosslinking agent, dimethoxymethane. A functionalized thioether bonded structure (C-S-C) in the SC composite could be disassembled using voltage scissors at a low range of 0.01–0.50 V vs. Na/Na$^+$. The insolvable sodium polysulfides were successfully entrapped in the carbon defects of SC composite and, thus, could avoid the degradation of soluble polysulfides. Despite their expedited reversibility, their practicality was impeded by a limited amount of loadings which caused a roadblock in constructing a high-energy sodium storage system.

Polymer-based composites, particularly polyacrylonitrile (PAN), have also been explored numerous times with great achievements in Li-S batteries owing to their robust structure and flexibility that enables the cathode composite to withstand volumetric fluctuations [35,119,120]. Accordingly, Kim and co-workers synthesized a singular flexible sulfurized polyacrylonitrile (SPAN) cathode, showing high bendability up to 180°. On cycling at 0.01 C, the electrode maintained over 266 mAh g^{-1} at the end of 200 cycles. As an inexpensive material, PAN with some modifications can be considered as having huge potential with paramount interest for the practicality of SPAN web material. Later, Li and co-workers reported a pyrolyzed polyacrylonitrile in combination with selenium sulfide (pPAN/SeS$_2$) with an active weight of 63 wt.% for sodium storage applications. The 1D framework of pPAN/SeS$_2$ demonstrated a high initial capacity and delivered a capacity of 800 mAh g^{-1} over 400 cycles at 1 A g^{-1} [121].

Attributing to a robust architecture, a new type of sulfur co-polymer-(CS-90) was fabricated, resulting in incorporating a uniform distribution of sulfur with maximum utilization of sulfur. Due to in situ formation and the beneficial effect of strong polymer architecture, minimal aggregation of decomposition product was observed with the reduced dissolution of intermediate sulfur moieties. Small organosulfide units of CS-90 co-polymer assisted with high conductivity of rGO bestow the electrode with a Coulombic efficiency of ~99% [122].

Li et al. designed a polymer-bound sulfurized polyacrylonitrile (S@pPAN) and chemically altered the approach with 4 mol % of tellurium dopant, resulting in a composite [123]. As a eutectic accelerator, Tellurium showed escalated electrochemical kinetics and boosted reversibility while inducing high conductivity in the covalently fixed composite (i.e., ~1.6 times of pPAN) (Figure 3.6b). Under different conditions, the composite was tested for its electrochemical performance, and the composite cathode exhibited fast reversible kinetics due to tellurium doping. The resultant cathode could attain a low attenuation rate of 0.015% per cycle at 0.5 A/ for over a long-term cycle life of 600 cycles owing to an enhanced sodium ion diffusion coefficient.

To date, different covalent bonded composites have made a breakthrough in developing RT Na-S battery systems. Nevertheless, covalently fixed composites are challenged with ineluctable damage in their interconnected structures due to swelling and polysulfide dissolution. Besides, they are also impeded by low sulfur content and discharge potential. However, it is anticipated that new possibilities will open up to end up with a high-end cathode system to expedite the commercialization of an RT Na-S battery technology

3.3.4 Hybrid S-host

Another possible strategy to mitigate the dissolution effect is combining the benefits of physical, chemical, or covalently bonded systems. The realized loadings for sulfur cathode are not yet satisfactory. As a result, many researchers have proposed new hybrid structures to meet the challenges in the RT Na-S battery chemistry.

Because of the above-said challenge, Hwang et al. [125] proposed a high-loaded sulfur cathode (6.4 mg cm^{-2}) of porous 3D interconnected carbon fiber network (CFC) with high S-utilization and content. They were successful in locking up the sulfur moieties inside the shell of a CFC-coated "redox-active polar shell," constituting a "bifunctional sheath of Fe(CN)$_6^{4-}$ doped polypyrrole film (FC-PPy)." Moreover, cation reservoir-like Fe(CN)$_6^{4-}$ doped in polypyrrole film allows accessibility to ions to boost the reversible kinetic reactions. In addition to the strong adsorptive capacity of Fe(CN)$_6^{4-}$ redox mediators, CFC/S@FC-PPy hybrid cathode illustrated a low UV peak intensity of 0.09 in sharp contrast to 0.55 and 0.24 of CFC/S and CFC/S@PPy, respectively, which illustrates the adsorption affinity toward polysulfides for different electrodes. The cathode demonstrated a phenomenal storage capacity of 1071 mAh g^{-1} with a retention of 72.8% when cycled for 200 cycles. Likewise, another group studied a hybrid freestanding structure of 3D-reinforced reduced graphene oxide (rGO) in combination with mixed-valence ultrafine manganese oxide nanocrystals (Mn$_x$O$_y$) [124]. To coalesce, the composite was mixed with sodium alginate/polyaniline (SA/PANI) adhesive matrix to serve as the host to form a sulfur cathode. The conductive matrix develops interparticle adhesion to ameliorate the overall facilitation of ions and electrons and forms a network to contain 2.05 mg cm^{-2} of sulfur. Mn$_x$O$_y$ catalytically helps reduce manganese from an oxidation state of +4 to +3 during discharge. The kinetic redox reaction between Mn$_x$O$_y$ and sodium polysulfides is schematically shown in Figure 3.6c.

Despite the challenges faced in a sulfur cathode, a tremendous breakthrough has been made in the past few years to propel the development of high-capacity utilization with increased S-loadings. Yet, the quest for a compatible sulfur cathode to overcome specific scientific and engineering challenges continues to access the practical aspect of the RT Na-S battery technology.

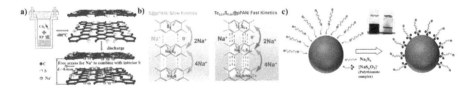

FIGURE 3.6 (a) Schematic presentation of covalent C/S composite as a capacity sponsor during sodiation, Adapted and reproduced with permission from Ref. [119]. Copyright © 2020 American Chemical Society. (b) Comparison schematic showing the reaction pathway for Te0.04S0.96@pPAN and S@pPAN composites, Adapted and reproduced with permission from Ref. [125]. Copyright © 2019 American Chemical Society. (c) A scheme presenting the surface redox reaction between MnxOy and Na2S6. Adapted and reproduced with permission from Ref. [126]. Copyright © 2019 American Chemical Society.

3.4 IMPORTANCE OF ELECTROLYTE/SULFUR (E/S) RATIO IN DEVELOPING A STABLE SULFUR CATHODE

Electrolyte significantly impacts the electrochemistry and kinetic reaction of the metal-sulfur battery [126–128]. One of the deciding factors that control the facilitation of ions/electrons within the cell is the conductivity of the electrolyte and considering the impact on the dissolution of polysulfides into the bargain. Nonetheless, taking into account the polysulfide's dissolution process and formation of solid electrolyte interface (SEI) [129,130], the depletion of electrolytes seems inevitable [131]. Notably, the sulfur distribution and utilization are highly dependent on the interaction between the cathode and the electrolyte system.

One of the main cathode design parameters, i.e., electrolyte/sulfur (E/S) ratio, plays an important role in deciding the performance of the battery [132]. Of late, the E/S ratio has been considered a key design parameter as it is a function of both the electrolyte's viscosity and polysulfide dissolution. The attainable specific energy density is low in metal-sulfur batteries, mainly due to the high E/S ratio. Therefore, by increasing the E/S ratio, the overall electrolyte causes an increase in the weight of the cell, which eventually decreases the energy density of the cell [133]. As a result, a lower E/S ratio is recommended for a higher energy-density system. However, a very low E/S ratio causes severe cyclability and degraded capacity challenges due to a lower amount of electrolyte impeding the facilitation of ions within the system. Therefore, E/S plays a critical role as a lever in controlling the energy density of the cell and maintaining the capacity within the cell [134]. Based on the contradiction, an optimum level of E/S ratio needs to be maintained considering the capacity of the cell without compromising the energy density of the system. The studies on the E/S ratio in RT Na-S batteries have rarely been put forward and have not gained enough attention. Studies have been conducted on Li-S cells based on the effects of the E/S ratio on the electrochemical performance of the battery. To date, much research has been focused on maintaining a high sulfur utilization and cyclability with an E/S ratio ^{3}of 10 µL mg^{-1} [135,136]. For instance, Zheng and co-workers investigated the importance of a high E/S ratio >20 µL mg^{-1} as an optimized ratio to extract the maximum capacity with minimized shuttle phenomena. On the contrary, fewer studies have been formulated on E/S ratio <4 µL mg^{-1} due to inadequate wetting of the electrode surface in addition to continuous degradation of electrolytes during cycling. Very little research has been put forward toward optimizing an accurate lower level of E/S ratio while achieving an improved capacity and cycling performance [132]. An optimized E/S ratio can differ from various cell chemistry and electrolyte technology. To achieve a 50% or higher energy density, the E/S can be maintained between 2 and 5 mL mg^{-1} [137–139].

It is to be noted that the percentage in volumetric change can be considered as a function of sulfur loading and E/S ratio. As an elucidation, Guo and Fu intensively studied the effect of the E/S ratio on the Li-S system [140]. Their findings state that as the E/S ratio augments from 2 to 10, the percentage of volumetric change diminishes, and the dependency on the E/S ratio reduces accordingly. Also, as the loading of sulfur increases from 2 to 4 mg cm^{-2}, the volume change increases and then flattens as the sulfur loading augments from 4 mg cm^{-2} and above. An explanation of the statement is illustrated, which depicts the percentage of volumetric change for sulfur loading and the E/S ratio about the volume of a cell taken initially. Accordingly, in-depth studies were further made on the E/S ratio by introducing another parameter, 'average mass density,' to evaluate the specific energy density of the cell. They have mentioned that the average mass density is a function of the E/S ratio and is based on the assumption that the sulfur reaction occurs in an excess electrolyte. Considering the densities of all the sulfur components, the average mass density is expected to decline with an increase in the E/S ratio. It is to be noted that the Li-ion batteries show an average mass density of 3.6 to 5 g cm^{-3} based on metal-oxide cathodes. As the E/S ratio varies between 4 and 10 μL mg^{-1}, the average mass density decreases to up to 1.15 g cm^{-3} of the liquid electrolyte in a Li-S battery, which is lower than the commercial Li-ion batteries. Based on 60% of S-utilization (i.e., 1000 mAh g^{-1}), they have analyzed a linear relationship between the average mass density and energy densities (specific and volumetric) of S-cathode, including liquid electrolyte. Corresponding to an E/S ratio of 4 μL mg^{-1}, the study shows a specific and volumetric energy density of 400 Wh kg^{-1} and 500 Wh L^{-1} when the average mass density of the cathode is 1.24 g cm^{-3}. On that account, new approaches and designs must be put together to focus on optimizing the E/S ratio below 5 μL mg^{-1} without compromising the sulfur utilization, capacity, and cycle life of the cell [141].

3.5 STRATEGIES TO DEVELOP HIGH-LOADING SULFUR CATHODES

As a benchmark, Pope et al. [142] reported "energy density as a function of sulfur loadings" as analyzed in Li-S cathode chemistries. Their studies have analyzed that the S-loadings should be >2 mg cm^{-2} to achieve a high specific energy of ~400 Wh kg^{-1} in a Li-S battery. Similarly, Liu and co-workers extensively investigated enhancing the volumetric energy density in a Li-S battery [143]. The key to increasing the specific density is incorporating a high sulfur percentage in the cathode composite. On that note, they have put forward certain principles to achieve a desirable S-cathode with high values of energy densities, such as (i) high S-content of >80 wt.% and (ii) S-loading greater than 4 mg cm^{-2} [144–146].

Generally, in a sulfur cathode composite, the sulfur fraction is maintained at 50%–80% by weight, binder composition from 3 to 10 wt.% or higher, and a suitable host at around 15%–20% by weight [147]. Therefore, the amount of sulfur content in the total weight fraction of the active material (i.e., approximately 90 wt.%) is relatively low. Addressing these issues, various strategies have been put forward to load a high amount of sulfur onto the electrode materials for practical applications. Table 3.2 provides an overview of multiple S-hosts cathodes based on different synthesis routes for RT Na-S battery technology.

3.5.1 SYNTHESIS TECHNIQUES

3.5.1.1 Physical mixing

To load sulfur into the conductive mixture, physical mixing has been used conventionally over the years, and such type depends purely on the nature of conductive hosts [148,149]. Various mixing approaches and physical mixing devices have been explored to fabricate good-performance electrode materials, for example, magnetic stirrers, ball mills, and ultrasonic devices [150–152]. This technique is energy-saving, simple, and can be utilized in large-scale applications.

3.5.1.2 Melt diffusion technique

A commonly used sulfur-loading approach is the melt diffusion strategy based on capillary adsorption, which involves high heating of sulfur-based components at around 150°C–175°C. As the viscosity of sulfur is low, the sulfur is allowed to melt on the application of heat and is integrated into the conductive materials [153–156]. Despite its application on a large scale, this technique results in low porous sulfur-based electrodes, which impedes the flow of ions/electrons into the mixture and the use of expensive and toxic solvents [157,158].

3.5.1.3 Vapor phase infiltration technique

In the sulfur vapor infiltration technique, the sulfur reservoir is placed below and separately from the conductive components in a reaction chamber. The components inside the enclosure can heat for a few hours at constant, elevated heating of ~175°C under vacuum. The vaporized sulfur is then allowed to infiltrate inside the pores and voids of the host matrix [159–161]. Although this can be a unique method for various systems and applications, this approach can consume a certain amount of energy [162].

3.5.1.4 Dissolution/re-crystallization technique

Nonetheless, several approaches result in sulfur agglomeration with a poor affinity toward sulfur products. To get more control over the interfacial interaction among the particles for better loading of sulfur, another approach called sulfur dissolution/re-crystallization came into light [163]. This technique is accomplished by choosing a specific solvent of suitable solubility to allow sulfur to recrystallize on the host and avoid any unconstrained self-recrystallization. First, the host particles and sulfur are physically mixed with magnetic stirring or ultrasonic devices. On the one hand, the dissolution/precipitation is reached in the solvent. On the other hand, the conductive hosts allow the absorption and stabilization of sulfur and act as the base for the crystal nucleus for the crystallization of sulfur. This concomitant dissolution and re-crystallization of sulfur allow a solid-liquid-solid phase transition, allowing a coated layer of sulfur on the host [164].

3.5.1.5 Direct drop-wise technique

Encapsulation approaches can be difficult for processing electrodes for large-scale flow production because of complex steps. Therefore, to promote colossal scale production of sulfur cathode without any complicated steps for pre-processing, an approach of loading sulfur via direct dropping of sulfur-based liquid into conductive hosts has been brought forward. Pre-processing steps such as heat treatment, sealing, or wrapping are redundant. Thus, this direct drip method can obtain a highly ordered sulfur composite [165].

3.5.1.6 Chemical methods

Due to the limited or no solubility of sulfur in the water, it causes issues. The sulfur-containing compounds such as sodium thiosulfate, sodium sulfide, sulfo-urea, and others are being used, which can be dissolved in various solvents. Based on the chemical properties of components, different methods such as chemical or electrochemical deposition, copolymerization technique, and oxidation method route can be used to develop high-loaded cathodes [150,166–168].

3.5.2 COATING TECHNIQUES

A conventional wet slurry method of fabricating the cathode is to take the sulfur, binder, and a conductive material (carbonaceous-based compounds, dopants, metal oxides, and others) and mix the slurry with a solvent via ball-milling. The follow-up ball-milling is done to enhance the contact area among the component and augment the migration of ions and electrons to improve the kinetic reactions [147]. Following this, a traditional blade-casting method is utilized to cast the mixture onto an aluminum current collector, directly heated in a vacuum to remove any solvent from the mixture, and calendered to form the electrode with desirable thickness [178,179]. However, such a traditional method of cathode fabrication suffers from poor porosity, low electrolyte infiltration into the matrix, and low utilization of sulfur [180]. Accordingly, new techniques to tailor the sulfur cathodes must be created upon surmounting these challenges.

3.5.2.1 Solvent-free electrostatic spray deposition

Solvent-free electrostatic spray deposition (ESD) or high-pressure deposition as an alternative to wet slurry processing was initially developed by Maxwell and further explored by various researchers [181]. Because of the large amount of organic solvent utilized in the conventional wet method, some of which are costly and toxic, green solvent-free dry coating approaches have been proposed. ESD does not involve thermal drying or organic solvents, making it a reliable option with less wastage of energy and avoiding eliminating volatile components. On application of high voltage, the powdered form of a composite material consisting of sulfur, conductive materials, and binder is mixed and placed on a current collector and is allowed to vaporize. Finally, the materials are calendered to attain an electrode with suitable porosity and thickness. Another solvent-free technique of hot pressing and ambient temperature pressing is in the trend to fabricate electrodes for all-solid-state batteries where the powdered material is hot pressed at ~ 100°C overnight and then cold-pressed to achieve an electrode with the desired thickness. Although depending on the pressure applied during the compression, the electrochemical performance of the electrode can vary [182–184].

3.5.2.2 Phase inversion

Phase inversion is another universal membrane processing technique for achieving high-loaded electrode composites [185,186]. Given that the electrode materials tend to peel off on increasing sulfur loading, resulting in a limited network for electron/ion transportation and electrolyte accessibility, the phase inversion method is known for its construction in a hierarchical porous-interconnected matrix of polymer skeleton [187–189]. During this process, the slurry coated onto a current collector is dipped into an aqueous (water) or non-aqueous (alcohol) coagulation bath. As a result, solvent/non-solvent exchange occurs. Due to the van der Waals force of interaction, three continuous phases are formed during the phase inversion process: (i) sulfur/carbon phase for enhancement of electron migration, (ii) polymer phase to bind and wrap the active materials, and (iii) pore phase to supply sufficient ions and allow ease of electrolyte permeation. The tri-continuous matrix can allow a highly conductive path for ions and electrons even at higher sulfur loadings [190].

3.5.2.2 3D printing

Another trailblazing cathode fabrication technique is the 3D printing technique which comes with several advantages; 3D printing opens new avenues that allow the fabrication of complex designs, enables accurate controlling of thickness, porosity, and stability, and provides a safe-green operation [191,192]. Low-temperature direct drying, material extrusion-type, directed energy deposition, and others are some developed 3D printing techniques used for cathode fabrication. As an advanced fabrication technique, 3D printing helps print 3D architecture based on digitally controlled reactant materials and solvent ink-based materials to form layered-by-layered 2D objects to give rise to a 3D electrode model finally. 3D printing can open new opportunities and light up promising routes to fabricate electrodes with high surface area and highly loaded materials with remarkable electrochemical performance and complex architectures [181].

As can be seen in Table 3.2, various strategies have been followed by different researchers to load high amounts of sulfur. However, with the increase in sulfur loadings, the system faces formidable challenges, such as capacity decay or an increase in polarization. As the thickness increases, the active material loses its contact electrochemically, and the electrode system is met with cracks and peeling-off issues. With a further increase in sulfur loadings, the system experiences hindrance in the uptake of ions and further resistivity drop. Increasing the thickness or sulfur-loading in a sodium-sulfur system is of unprecedented importance. In this regard, in-depth research needs to be envisaged to burgeon new strategies or chemistry areas in an electrode material to increase the loadings and ensure the maximum possible energy requirements.

TABLE 3.2
Performance data for various S-hosts cathode based on different synthesis routes for RT Na-S battery technology

Cathode composite	Synthesis route	Active mass loading (mg/cm^2)	Electrolyte system	Initial capacity (mAh g^{-1})	Capacity retained (mAh g^{-1}) after [n cycles]	C-rate or current density
Physically confined sulfur composite						
Carbon fiber cloth@sulfur [47]	Solution impregnation strategy	2$_{(S)}$	1.5 M NaClO$_4$ & 1.0 M NaNO$_3$ in tetraethylene glycol dimethyl ether	390	120 [300]	0.1 C
Nanocarbon-promoted comm@sulfur [51]	Physical mixing	0.8–2.0$_{(S)}$	Gel polymer electrolyte: 1 M NaPF$_6$/diglyme	713	700 [200]	0.2 C
Sugar-derived microporous carbon@sulfur [59]	Isothermal vapor phase infiltration	–	1 M NaPF$_6$ & 0.25 M NaNO$_3$ in TEGDME	~410	306 [1500]	1 C
Mesoporous carbon hollow nanospheres@sulfur [55]	Sol-gel method	3.2–4.1	1.0 M NaClO$_4$ PC/EC with 5 wt.% FEC	1215	292 [200]	0.1 A g^{-1}
Ultramicroporous carbon@sulfur [64]	Single step polymer carbonization	3	1 M NaClO$_4$ in EC/DEC	–	392 [200]	1 C
Activated ultramicroporous coffee carbon@sulfur [52]	Annealing	1$_{(S)}$	1.0 M NaClO$_4$ PC/EC with 2 wt.% FEC	1492	1110 [400]	0.1 C
Porous carbon microspheres@sulfur [53]	Chemical vapor deposition	–	1.0 M NaClO$_4$ PC/EC with 5 wt.% FEC	1100	290 [350]	0.1 A g^{-1}
Carbon hollow nanobubbles@Porous carbon nanofibers@sulfur [54]	Electrospinning technique	1.4$_{(S)}$	1.0 M NaClO$_4$ PC/EC with 5 wt.% FEC	~300	256 [400]	2 C

(Continued)

TABLE 3.2 (CONTINUED)

Cathode composite	Synthesis route	Active mass loading (mg/cm^2)	Electrolyte system	Initial capacity (mAh g^{-1})	Capacity retained (mAh g^{-1}) after [n cycles]	C-rate or current density
Chemically confined sulfur composite						
N-cZIF-8@sulfur [169]	Carbonization	0.7–0.9	1 M NaClO$_4$ in TEGDME	873	500 [250]	0.2 C
N, S doped-hierarchical porous carbon [79]	Customized roll-to-roll line strategies	1.0–1.1	Na$_2$S in TEGDME	~430	378 [350]	0.23 A g^{-1}
Nitrogen-doped graphene nanosheets@sulfur [80]	Chemical reaction deposition strategy & low-temperature heat treatment	–	0.8 M NaClO$_4$ in EC/DMC	~110	48 [300]	0.1 C
MOF-derived S,N-doped porous carbon [170]	Vapor-infiltration method	0.9–1.5	1 M NaClO$_4$ in PC with 5 wt.% FEC	1000	680 [500]	1 A g^{-1}
Fe-hollow carbon nanospheres@sulfur [85]	Sol-gel process	0.9–1.1	1 M NaClO$_4$ in PC/EC with wt.% FEC	1023	394 [1000]	0.1 A g^{-1}
Atomic cobalt-hollow carbon nanospheres@sulfur [86]	Sol-gel process	5	1 M NaClO$_4$ in PC/EC with 5 wt.% FEC	1081	508 [600]	0.1 A g^{-1}
Nickel-Nitrogen-doped carbon fibers@sulfur [87]	Electrostatic spinning process	0.5–0.7	1 M NaClO$_4$ in TEGDME	~431	~233 [270]	1 C
Nitrogen-doped carbon nanospheres loaded with Fe@sulfur [88]	Carbothermal reduction followed by calcination process	–	2 M NaTFSI in PC/FEC	>350	180 [200]	1 A g^{-1}
Carbon wrapped cobalt nanoparticles on graphene aerogel@sulfur [89]	Thermal treatment	–	1 M NaClO$_4$ in TEGDME	572.8	286.5 [1000]	5 C

(*Continued*)

TABLE 3.2 (CONTINUED)

Cathode composite	Synthesis route	Active mass loading (mg/cm²)	Electrolyte system	Initial capacity (mAh g⁻¹)	Capacity retained (mAh g⁻¹) after [n cycles]	C-rate or current density
Gold nanodots on hierarchical N-doped carbon microspheres@ sulfur [90]	HCl etching process	–	1 M NaClO$_4$ in PC with 5 wt.% FEC	1967	701 [110]	0.1 A g⁻¹
CNF-L@Co/@ sulfur [91]	Elctrospinning and solution method	1–1.2$_{(S)}$	1 M NaClO$_4$ in EC/DEC	736.8	538 [150]	0.5 C
Ni-MOF-2D@ sulfur [28]	Hydrothermal strategy	1	Na$_2$S & S in EC/PC	600	347 [1000]	1 C
CoP/Co-N-porous carbon nanotube hollow cages@ sulfur [92]	Sequential carbonization-oxidation-phosphidation	1.7	1 M NaClO$_4$ in EC/PC with 3 wt.% FEC	>400	448 [700]	1 A g⁻¹
C/S/BaTiO$_3$@ titanium dioxide [100]	Electrospinning technique and atomic layer deposition	1.2–1.4	1 M NaClO$_4$ in EC/DEC	952	524.8 [1400]	1 A g⁻¹
TiN-TiO2@ MCCFs@sulfur [101]	Electrospinning and nitridation process	1.08$_{(S)}$	1 M NaClO$_4$ in EC/PC with 5 wt.% FEC	~500	257.1 [1000]	5 A g⁻¹
Molybdenum sulfide [104]	Chemical oxidation	1.5–2	1 M NaPF$_6$ in DEC/FEC	400	311 [100]	0.2 A g⁻¹
Core-shell ZnS and CoS2@ sulfur [107]	Carbonization and sol-gel process	2	1 M NaClO$_4$ in DEC/EC with 5 wt.% FEC	>1900	570 [1000]	0.2 A g⁻¹
MOF-based CoS2/N-carbon@sulfur [93]	Drop-wise technique and heat treatment	1.2–1.5	1 M NaSO$_3$CF$_3$ in diethylene glycol dimethyl ether	~630	403 [1000]	1 A g⁻¹
CoS$_2$C@sulfur [94]	Simple reflux method	4.4	1 M NaClO$_4$ in TEGDME	1347	675 [800]	0.5 C

(*Continued*)

TABLE 3.2 (CONTINUED)

Cathode composite	Synthesis route	Active mass loading (mg/cm^2)	Electrolyte system	Initial capacity (mAh g^{-1})	Capacity retained (mAh g^{-1}) after [n cycles]	C-rate or current density
N-doped carbon matrices decorated by polar cobalt sulfide (CoS$_2$) and single-atom Co [171]	Two-step high-temperature annealing	1.6	1 M NaClO4 in EC/PC with 5 wt.% FEC	>1150	357 [5000]	5 A g^{-1}
Tungsten nanoparticles embedded in nitrogen-doped graphene [172]	In situ pyrolysis	1	1.0 M NaClO$_4$ PC/EC with 3 wt.% FEC	~850	398 [1000]	1 C
Tessellated N-doped carbon/CoSe$_2$ [173]	Simple deposition–selenization	2.15	1 M NaClO$_4$ in EC/DEC	>800	470.3 [500]	1 C
Cobalt sulfide selenide hetero-structure in multichannel CNFs [174]	Carbonization followed by in situ sulfuration and selenization.	–	2 M NaTFSI in PC & FEC	~1800	749 [200]	1 A g^{-1}
Covalently fixed sulfur composite						
Covalent Sulfur-carbon complex (SC-BDSA) [116]	Physical mixing and annealing	3	1 M NaClO$_4$ in EC/PC	696	452 [1000]	2.5 A g^{-1}
Covalent Sulfur into carbon matrix (SC) [117]	Wet solvothermal reaction	6.3	1 M NaClO$_4$ in EC/DEC with 5 wt.% FEC	1063.7	811.4 [950]	1.6 C
Thioether bond functionalized carbon@sulfur [118]	Carbonization	0.9-1-1	NaClO$_4$ in PC with 5 vol% C3H3FO3	~900	330 [800]	1 A g^{-1}
SPAN webs [175]	Pyrolysis	0.73	1 M NaPF$_6$ in EC/DEC	342	266 [200]	0.1 C

(*Continued*)

TABLE 3.2 (CONTINUED)

Cathode composite	Synthesis route	Active mass loading (mg/cm²)	Electrolyte system	Initial capacity (mAh g⁻¹)	Capacity retained (mAh g⁻¹) after [n cycles]	C-rate or current density
Pyrolyzed PAN/Selenium disulfide [121]	Electrospinning and heat treatment	2	1 M NaClO$_4$ in EC and DMC with 5 wt.% FEC	1043	800 [400]	1 A g⁻¹
Carbon-Sulfur90/rGO [122]	Thermal ring-opening polarization	2.14$_{(S)}$	1 M NaClO$_4$ and 0.2 M NaNO$_3$ in TEGDME	~500	285 [100]	1 A g⁻¹
Tellurium-doped/SPAN [123]	Heat treatment method	1.0–1.2	1 M NaClO$_4$ in EC and DMC with 10 wt.% FEC	~1500	970 [600]	0.5 A g⁻¹
SPAN with PAA binder [176]	Heat treatment	0.5–0.7$_{(S)}$	1 M NaClO$_4$ in EC and DMC with 40 mM SnCl$_2$ and 8 vol.% FEC	1195	1000 [1000]	0.5 C
Polydopamine derived N-doped carbon layer coated onto graphene [177]	Vapor-infiltration	1.2–1.5	1.0 M NaClO$_4$ PC with 5 wt.% FEC	~590	356 [1000]	1 C
Hybrid sulfur composite						
Carbon fiber cloth/S@Fe(CN)$_6^{4-}$ doped polypyrrole film [125]	Thermal alkali activation and interfacial polarization strategy	2.2–2.8	1 M NaFSI in TEGDME with NaNO$_3$ additive	~1071	700 [200]	0.2 A g⁻¹
rGO/S/Mn$_x$O$_y$@SA–PANI [124]	Chemical precipitation, heat treatment and vacuum filtration	2.05$_{(S)}$	1 M NaClO$_4$ in TEGDME with 0.1 M NaNO$_3$ additive	>700	535 [50]	0.2 A g⁻¹

3.6 FUTURE PROSPECTS

With the evolution of progress toward sodium-sulfur battery technology, the kinetic behavior of polysulfides constitutes an increasingly significant roadblock in battery systems [193]. Despite the general mechanistic understanding of a sodium-sulfur battery is known, however, demonstrating a strong sense of the mechanism of polysulfide species is the key to the enhancement in the advancement of high-energy battery chemistry.

Various prospects can be applied to achieve a high-end battery based on advanced characterization techniques and computational models. Trailblazing practices have emerged from progress made on Li-S batteries. The following developments on Li-S systems can likely bring optimistic effects on the RT Na-S batteries.

3.6.1 ADVANCED CHARACTERIZATION TECHNIQUES

Understanding the importance of meticulous research on the discharge mechanisms of sulfur cathode on the performance of the RT Na-S battery is crucial. Various parameters have been quantitatively explored through traditional particle and cell-level approaches. Complex phenomena and phase transitions of sulfur moieties follow in a cathode, making it difficult for researchers to conduct in-depth studies. To make this viable, it is imperative to develop up-to-date techniques to understand the effect of inherent properties and mechanisms and their correlation to electrochemical performances. Accordingly, a swathe of advanced in situ characterization technologies has been well-explored for the sulfur cathode in Li-S battery technology [194]. It is worth noting that these techniques allow real-time studies during the chemical redox reaction and analyzes the intermediate processes under real operating conditions. In contrast to the ongoing ex situ techniques, which can only give the details before and after the electrochemical studies [195], in situ technologies deliver information on (i) the inherent relationship between the transition of components and its further electrochemical information, (ii) the transition of volume and phases, (iii) morphology fluctuations, and (iv) real-time intuitive studies on the electrode-electrolyte interface. In addition, they pay heed to dynamic characteristics and their kinetics without the influence of air or humidity [196,197].

For example, deciphering the mechanistic reaction occurring in a Li-S cell using in situ or operando synchrotron-based technologies: X-ray diffraction (XRD), X-ray absorption spectroscopy (XAS), and X-ray microscopy (XRM), for example [198]. The above techniques allow visualization of the evolution of chemicals and structures under real-operating conditions. For instance, synchrotron-based XRD is a powerful tool to analyze the changes in crystallographic changes, including phase, strain, and lattice parameters. Time-resolved synchrotron-based XRD (TR-XRD) has recently been implemented to study the dynamic nature of materials under real-operating conditions during the fast cycling process. In situ XAS is another technique for observing the changes in an electronic structure and local structural changes at both atomic and molecular levels. Direct information on the chemical composition, morphology, and microstructure can be regulated using different synchrotron-based XRM, such as X-ray fluorescence microscopy (XRF), transmission X-ray microscopy (TXM), and X-ray tomographic microscopy (XTM).

The dissolution of polysulfides has been one of the critical challenges that obstruct the feasibility of the sulfur cathode. Completely comprehending the reaction mechanism of the shuttle effect in sulfur cathode remains ambiguous. As a result, various studies have been put forward on synchrotron-based techniques to study the evolution of polysulfides during charge/discharge cycles. Accordingly, a study based on in situ XRD allowed direct visualization of polysulfides during cycling and monitored their absorption and development on the surface of a glass-fiber separator [199]. They observed the slow aging of the cell by the addition of silicon dioxide as an additive in the electrolyte, which led to an increase in the charge by relatively 25% as compared to the reference electrolyte system without the additive. The absorption of polysulfides could be observed by faster scavenging of the species, resulting in a buffered redox shuttle effect. Recently, in situ confocal Raman microscopy has been implemented to study the kinetics of the multi-step conversion redox pathways [200]. They also provide an in-depth investigation into sulfur deposition and the generation or evolution of lithium polysulfides during cell operation.

The operando studies allowed detailed investigation of the reactants and intermediates, which directed a relationship of the reaction kinetic rate with the polysulfides' concentration and sulfur's conductivity. It is to be noted that the operando quantification also allowed the characterization of the 'potential-dependent rates' during the redox process. Similarly, Jia et al. [201] reported on an in situ XAS to investigate sulfur species' migration and monitor the dynamic speciation at the interface under actual operating conditions. To mitigate the shuttling effect, they employed an electrocatalytic composite of bismuth sulfide and oxide in a carbon scaffold to mitigate the polysulfide dissolution. The bifunctional layer successfully suppresses the polysulfide dissolution as captured using the operando XAS. A high spatial resolution can be complicated at a nanoscale level under real-time conditions. However, atomic force microscopy-based scanning electrochemical microscopy (AFM–SECM) makes this possible [202]. During oxidation reactions, the in situ phase mappings of solid state Li_2S based on electrochemical and alternating current depicted the presence of insulating and conducting parts. As a result, the insulating component (Li_2S) reacted with the intermediate polysulfides, whereas the conducting component dissolved into the electrolyte system during charging. However, at even higher oxidation potentials, the polysulfides formed reacted to form more insulating products, accumulating over time, resulting in loss of active material and, finally, capacity fade. AFM–SECM as a characteristic tool provided new insight into the interdependence of Li_2S topography-oxidative behavior on the carbon matrix.

3.6.2 COMPUTATIONAL MODELS

Researchers need to put strenuous effort into fully understanding the complex phenomena in the RT Na-S battery based on the complex mechanisms that follow during conversion reactions in a sulfur cathode and the absence of in situ characterization techniques. The microscale representation level of modeling gives us the fundamental phenomena occurring at the particle level and elucidates the microkinetics rather mechanistically. Also, macroscale modeling provides us with a perception of the design and optimization at the full cell level. Beyond that, to comprehend the fundamental interactions at the atomistic scale, various models such as Density functional theory (DFT), ab initio molecular dynamics (AIMD), kinetic Monte Carlo (KMC), and others can be performed [203]. Based on the simulations and first principle calculations, the various state-of-the-art models can be utilized to study the fundamental molecular interactions and determine multiple parameters such as reaction pathways or information on the facilitation of ions-electrons or energies related to absorption and surface level.

Lin's work coupled experimental validation and computational modeling to design and optimize open ring sizes (ORS) on the carbon nanotubes-sulfur encapsulated composite when oxidized within a particularly narrow range [204]. With integrated DFT calculations, they optimized the ORS generated in the walls of CNT and calculated the barrier to the transportation of Li$^+$ ions. Further estimates were made based on the molecular sizes of solvent and polysulfides with the ORS to selectively allow the facilitation of ions and simultaneously obstruct the dissolution of polysulfides. In the following year, Zhang et al. [205] performed DFT calculations using the Vienna ab initio simulation package (VASP) to screen the various functionalized MXenes host ($Ti_3C_2T_2$) (T = NH, S, O, H, Se, Cl, F, Br, OH, and Te). Furthermore, a method-projector augmented wave (PAW) was used in the calculations to analyze the interactions between the valence electrons and ions, and another method Per-dew Burke Ernzerhof (PBE), to describe the exchange-correlation function of electrons. Based on the DFT calculations, the catalytic effect in addition to the anchoring abilities of $Ti_3C_2T_2$ on soluble and insoluble poysulfides were investigated. As a result, three findings could be analyzed using the modelling: (i) to trap the Li$^+$ ions, there should be sufficient interactions existing between the sulfur species and surface groups, (ii) possible formation of S-H bonds from hydrogen present in the MXene host to enhance the anchoring effect of non-polar sulfur and its moieties, and (iii) relation of the electronegativities of elements within the same group to the binding energy of polysulfides. Another study reveals modeling using DFT calculations to understand the anchoring of lithium polysulfides at the surfaces of various transition metal sulfide (SnS_2, NbS_2, MoS_2, TiS_2, WS_2, and VS_2) [206]. The study emphasizes how the geometric property of the surface regulates the adsorption energy of polysulfide species. They have set a geometric parameter as G_{score} (f{bond length, number of Li-atom interacting between the polysulfide species and the transition metal surface}). G_{score} can be used to predict the optimum binding energy zone for a practical electrocatalytic effect. Furthermore, the DFT simulations made predictions based on the formation of allotrope and repulsion energy of the pore wall [207]. DFT calculations were analyzed on a porous and an activated carbon host to study the energies of various allotropes of sulfur (cyclic S_8 and chain S_8, S_6, and S_4) in different sizes of micropores.

We believe combining advanced characteristic technologies with computational modeling can deeply scrutinize a sulfur cathode's various complex mechanistic reactions. It may provide a perceptive understanding of the transition of phase and morphology with the electrochemical performance of the RT Na-S battery.

RT Na-S batteries are in the developing phase, and various challenges are yet to be addressed for the cathode to bring it to the market in a commercial aspect. A vigorous study must be formulated to develop state-of-the-art strategies to optimize the E/S ratio, build a robust architecture, and increase the sulfur content and its utilization. Notably, a comprehensive study must be conducted to design a compatible sulfur host to serve the increased demands of energy-related industries.

REFERENCES

1. Kumar D, Rajouria SK, Kuhar SB, et al. Progress and Prospects of Sodium-Sulfur Batteries: A Review. *Solid State Ionics.* 2017;312:8–16.
2. Zhu J, Zou J, Cheng H, et al. High Energy Batteries Based on Sulfur Cathode. *Green Energy Environ.* 2019;4:345–359.
3. Sungjemmenla, Vineeth SK, Soni CB, et al. Understanding the Cathode–Electrolyte Interphase in Lithium-Ion Batteries. *Energy Technol.* 2022;10:1–12.
4. Eng AYS, Soni CB, Lum Y, et al. Theory-Guided Experimental Design in Battery Materials Research. *Sci Adv.* 2022;8.
5. Sungjemmenla, Soni CB, Kumar V. Recent Advances in Cathode Engineering to Enable Reversible Room-Temperature Aluminium–Sulfur Batteries. *Nanoscale Adv.* 2021;3:1569–1581.
6. Soni CB, Arya N, Sungjemmenla, et al. Microarchitectures of Carbon Nanotubes for Reversible Na Plating/Stripping Toward the Development of Room-Temperature Na–S Batteries. *Energy Technol.* 2022;10:2200742.
7. Soni CB, Sungjemmenla, Vineeth SK, et al. Patterned Interlayer Enables a Highly Stable and Reversible Sodium Metal Anode for Sodium-Metal Batteries. *Sustain Energy Fuels.* 2023;7:1908–1915.
8. Yu X, Manthiram A. A Progress Report on Metal–Sulfur Batteries. *Adv Funct Mater.* 2020;30:1–27.
9. Soni CB, Kumar V, Seh ZW. Guiding Uniform Sodium Deposition through Host Modification for Sodium Metal Batteries. *Batter Supercaps.* 2022;5:e202100207:1–8.
10. Nikiforidis G, van de Sanden MCM, Tsampas MN. High and Intermediate Temperature Sodium–Sulfur Batteries for Energy Storage: Development, Challenges and Perspectives. *RSC Adv.* 2019;9:5649–5673.
11. Kumar D, Kuhar SB, Kanchan DK. Room Temperature Sodium-Sulfur Batteries as Emerging Energy Source. *J Energy Storage.* 2018;18:133–148.
12. Sungjemmenla, Soni CB, Vineeth SK, et al. Unveiling the Physiochemical Aspects of the Matrix in Improving Sulfur-Loading for Room-Temperature Sodium–Sulfur Batteries. *Mater Adv.* 2021;2:4165–4189.
13. Li S, Jin B, Zhai X, et al. Review of Carbon Materials for Lithium-Sulfur Batteries. *ChemistrySelect.* 2018;3:2245–2260.
14. Wang DW, Zeng Q, Zhou G, et al. Carbon-Sulfur Composites for Li-S Batteries: Status and Prospects. *J Mater Chem A.* 2013;1:9382–9394.
15. Li Z, Huang Y, Yuan L, et al. Status and Prospects in Sulfur–Carbon Composites as Cathode Materials for Rechargeable Lithium–Sulfur Batteries. *Carbon.* 2015;92:41–63.
16. Wang J-G, Xie K, Wei B. Advanced Engineering of Nanostructured Carbons for Lithium–Sulfur Batteries. *Nano Energy.* 2015;15:413–444.
17. Wang M, Xia X, Zhong Y, et al. Porous Carbon Hosts for Lithium–Sulfur Batteries. *Chem – A Eur J.* 2019;25:3710–3725.
18. Sungjemmenla, Soni CB, Vineeth SK, et al. Exploration of the Unique Structural Chemistry of Sulfur Cathode for High-Energy Rechargeable Beyond-Li Batteries. *Adv Energy Sustain Res.* 2022;3:2100157.
19. Jishnu NS, Vineeth SK, Das A, et al. Electrospun PVdF and PVdF-co-HFP-Based Blend Polymer Electrolytes for Lithium Ion Batteries. In: Balakrishnan NTM, Prasanth R, editors. *Electrospinning for Advanced Energy Storage Applications.* 1st ed. Singapore: Springer Singapore; 2021. pp. 201–234.
20. Vineeth SK, Tebyetekerwa M, Liu H, et al. Progress in the Development of Solid-State Electrolytes for Reversible Room-Temperature Sodium-Sulfur Batteries. *Mater Adv.* 2022;2:6415–6440.
21. Fang R, Xu J, Wang DW. Covalent Fixing of Sulfur in Metal-Sulfur Batteries. *Energy Environ Sci.* 2020;13:432–471.

22 Ma L, Zhuang H, Lu Y, et al. Tethered Molecular Sorbents: Enabling Metal-Sulfur Battery Cathodes. *Adv Energy Mater.* 2014;4:1–9.
23 Yu X, Manthiram A. Ambient-Temperature Energy Storage with Polyvalent Metal–Sulfur Chemistry. *Small Methods.* 2017;1:1–11.
24 Chung SH, Manthiram A. Current Status and Future Prospects of Metal–Sulfur Batteries. *Adv Mater.* 2019;31:39–42.
25 Zeng F, Wang W, Wang A, et al. Multidimensional Polycation β-Cyclodextrin Polymer as an Effective Aqueous Binder for High Sulfur Loading Cathode in Lithium-Sulfur Batteries. *ACS Appl Mater Interfaces.* 2015;7:26257–26265.
26 Zhang Z, Kong LL, Liu S, et al. A High-Efficiency Sulfur/Carbon Composite Based on 3D Graphene Nanosheet@Carbon Nanotube Matrix as Cathode for Lithium–Sulfur Battery. *Adv Energy Mater.* 2017;7:1–12.
27 Zeng F, Wang A, Wang W, et al. Strategies of Constructing Stable and High Sulfur Loading Cathodes Based on the Blade-Casting Technique. *J Mater Chem A.* 2017;5:12879–12888.
28 Ye C, Jiao Y, Chao D, et al. Electron-State Confinement of Polysulfides for Highly Stable Sodium–Sulfur Batteries. *Adv Mater.* 2020;32:1907557.
29 Su Y. Challenges and Prospects of Lithium-Sulfur Batteries. *Acc Chem Res.* 2013;46:1125–1134.
30 Hofmann AF, Fronczek DN, Bessler WG. Mechanistic Modeling of Polysulfide Shuttle and Capacity Loss in Lithium–Sulfur Batteries. *J Power Sources.* 2014;259:300–310.
31 Yang T, Qian T, Liu J, et al. A New Type of Electrolyte System to Suppress Polysulfide Dissolution for Lithium-Sulfur Battery. *ACS Nano.* 2019;13:9067–9073.
32 Xiong S, Xie K, Diao Y, et al. Characterization of the Solid Electrolyte Interphase on Lithium Anode for Preventing the Shuttle Mechanism in Lithium-Slfur Batteries. *J Power Sources.* 2014;246:840–845.
33 Cheng XB, Huang JQ, Peng HJ, et al. Polysulfide Shuttle Control: Towards a Lithium-Sulfur Battery with Superior Capacity Performance up to 1000 Cycles by Matching the Sulfur/Electrolyte Loading. *J Power Sources.* 2014;253:263–268.
34 Deng C, Wang Z, Wang S, et al. Inhibition of Polysulfide Diffusion in Lithium-Sulfur Batteries: Mechanism and Improvement Strategies. *J Mater Chem A.* 2019;7: 12381–12413.
35 Wei S, Ma L, Hendrickson KE, et al. Metal-Sulfur Battery Cathodes Based on PAN-Sulfur Composites. *J Am Chem Soc.* 2015;137:12143–12152.
36 Zhao J, Zhao D, Li L, et al. Defect-Rich, Mesoporous Cobalt Sulfide Hexagonal Nanosheets as Superior Sulfur Hosts for High-Rate, Long-Cycle Rechargeable Lithium-Sulfur Batteries. *J Phys Chem C.* 2020;124:12259–12268.
37 Yu X, Deng J, Lv R, et al. A Compact 3D Interconnected Sulfur Cathode for High-Energy, High-Power and Long-Life Lithium-Sulfur Batteries. *Energy Storage Mater.* 2019;20:14–23.
38 Richter R, Häcker J, Zhao-Karger Z, et al. Degradation Effects in Metal-Sulfur Batteries. *ACS Appl Energy Mater.* 2021;4:2365–2376.
39 Zou Q, Lu Y. Liquid Electrolyte Design for Metal-Sulfur Batteries: Mechanistic Understanding and Perspective. *EcoMat.* 2021;3:1–13.
40 Ford HO, Doyle ES, He P, et al. Self-Discharge of Magnesium-Sulfur Batteries Leads to Active Material Loss and Poor Shelf Life. *Energy Environ Sci.* 2021;14:890–899.
41 Wang Y-X, Zhang B, Lai W, et al. Room-Temperature Sodium-Sulfur Batteries: A Comprehensive Review on Research Progress and Cell Chemistry. *Adv Energy Mater.* 2017;7:1602829.
42 Manthiram A, Fu Y, Chung S, et al. Rechargeable Lithium – Sulfur Batteries. *Chem Rev.* 2014;114:11751–11787.
43 Richter R, Häcker J, Zhao-Karger Z, et al. Insights into Self-Discharge of Lithium- And Magnesium-Sulfur Batteries. *ACS Appl Energy Mater.* 2020;3:8457–8474.

44 Zhou L, Yao L, Li S, et al. The Combination of Intercalation and Conversion Reactions to Improve the Volumetric Capacity of the Cathode in Li-S Batteries. *J Mater Chem A.* 2019;7:3618–3623.
45 Zhang J, Ye H, Yin Y, et al. Core-Shell Meso/Microporous Carbon Host for Sulfur Loading toward Applications in Lithium-Sulfur Batteries. *J Energy Chem.* 2014;23:308–314.
46 Wang Z, Xu X, Ji S, et al. Recent Progress of Flexible Sulfur Cathode Based on Carbon Host for Lithium-Sulfur Batteries. *J Mater Sci Technol.* 2020;55:56–72.
47 Lu Q, Wang X, Cao J, et al. Freestanding Carbon Fiber Cloth/Sulfur Composites for Flexible Room-Temperature Sodium-Sulfur Batteries. *Energy Storage Mater.* 2017;8:77–84.
48 Kim SI, Park W Il, Jung K, et al. An Innovative Electronically-Conducting Matrix of the Cathode for Sodium Sulfur Battery. *J Power Sources.* 2016;320:37–42.
49 Zhao J, Yang M, Yang N, et al. Hollow Micro-/Nanostructure Reviving Lithium-sulfur Batteries. *Chem Res Chinese Univ.* 2020;36:313–319.
50 Li Z, Wu H Bin, Lou XW. Rational Designs and Engineering of Hollow Micro-/Nanostructures as Sulfur Hosts for Advanced Lithium-Sulfur Batteries. *Energy Environ Sci.* 2016;9:3061–3070.
51 Hu X, Ni Y, Wang C, et al. Facile-Processed Nanocarbon-Promoted Sulfur Cathode for Highly Stable Sodium-Sulfur Batteries. *Cell Reports Phys Sci.* 2020;1:100015.
52 Guo Q, Li S, Liu X, et al. Ultrastable Sodium–Sulfur Batteries without Polysulfides Formation Using Slit Ultramicropore Carbon Carrier. *Adv Sci.* 2020;7:1–12.
53 Zhang L, Zhang B, Dou Y, et al. Self-Assembling Hollow Carbon Nanobeads into Double-Shell Microspheres as a Hierarchical Sulfur Host for Sustainable Room-Temperature Sodium-Sulfur Batteries. *ACS Appl Mater Interfaces.* 2018;10:20422–20428.
54 Xia G, Zhang L, Chen X, et al. Carbon Hollow Nanobubbles on Porous Carbon Nanofibers: An Ideal Host for High-Performance Sodium-Sulfur Batteries and Hydrogen Storage. *Energy Storage Mater.* 2018;14:314–323.
55 Wang Y-X, Yang J, Lai W, et al. Achieving High-Performance Room-Temperature Sodium–Sulfur Batteries with S@Interconnected Mesoporous Carbon Hollow Nanospheres. *J Am Chem Soc.* 2016;138:16576–16579.
56 Noh H, Choi S, Kim HG, et al. Size Tunable Zeolite-Templated Carbon as Microporous Sulfur Host for Lithium-Sulfur Batteries. *ChemElectroChem.* 2019;6:558–565.
57 Li M, Liu Z, Zhang Y, et al. Nitrogen-Doped Microporous Carbon with Narrow Pore Size Distribution as Sulfur Host to Encapsulate Small Sulfur Molecules for Highly Stable Lithium-Sulfur Batteries. *J Solid State Electrochem.* 2021;25:1293–1302.
58 Hu L, Lu Y, Li X, et al. Optimization of Microporous Carbon Structures for Lithium–Sulfur Battery Applications in Carbonate-Based Electrolyte. *Small.* 2017;13:1–10.
59 Carter R, Oakes L, Douglas A, et al. A Sugar-Derived Room-Temperature Sodium Sulfur Battery with Long Term Cycling Stability. *Nano Lett.* 2017;17:1863–1869.
60 Wei S, Xu S, Agrawral A, et al. A Stable Room-Temperature Sodium-Sulfur Battery. *Nat Commun.* 2016;7:11722.
61 Wang C, Cui J, Fang X, et al. Regulating the Deposition of Insoluble Sulfur Species for Room Temperature Sodium-Sulfur Batteries. *Chem Res Chinese Univ.* 2022;38:128–135.
62 Liao K, Chen S, Wei H, et al. Micropores of Pure Nanographite Spheres for Long Cycle Life and High-Rate Lithium-Sulfur Batteries. *J Mater Chem A.* 2018;6:23062–23070.
63 Zhang Y, Zong X, Zhan L, et al. Double-Shelled Hollow Carbon Sphere with Microporous Outer Shell towards High Performance Lithium-Sulfur Battery. *Electrochim Acta.* 2018;284:89–97.
64 Hu L, Lu Y, Zhang T, et al. Ultramicroporous Carbon through an Activation-Free Approach for Li-S and Na-S Batteries in Carbonate-Based Electrolyte. *ACS Appl Mater Interfaces.* 2017;9:13813–13818.

65 Xu J, Lawson T, Fan H, et al. Updated Metal Compounds (MOFs, S, OH, N, C) Used as Cathode Materials for Lithium–Sulfur Batteries. *Adv Energy Mater.* 2018;8:1–23.
66 Zhu Y, Wang S, Miao Z, et al. Novel Non-Carbon Sulfur Hosts Based on Strong Chemisorption for Lithium–Sulfur Batteries. *Small.* 2018;14:1–21.
67 Yan Z, Liang Y, Xiao J, et al. A High-Kinetics Sulfur Cathode with a Highly Efficient Mechanism for Superior Room-Temperature Na–S Batteries. *Adv Mater.* 2020; 32:1–10.
68 Wang L, Dong Z, Wang D, et al. Covalent Bond Glued Sulfur Nanosheet-Based Cathode Integration for Long-Cycle-Life Li-S Batteries. *Nano Lett.* 2013;13:6244–6250.
69 Hou TZ, Chen X, Peng HJ, et al. Design Principles for Heteroatom-Doped Nanocarbon to Achieve Strong Anchoring of Polysulfides for Lithium–Sulfur Batteries. *Small.* 2016;12:3283–3291.
70 Zhang J, You C, Wang J, et al. Confinement of Sulfur Species into Heteroatom-Doped, Porous Carbon Container for High Areal Capacity Cathode. *Chem Eng J.* 2019;368:340–349.
71 Wang J, Han W. A Review of Heteroatom Doped Materials for Advanced Lithium–Sulfur Batteries. *Adv Funct Mater.* 2022;32:2107166.
72 Song J, Xu T, Gordin ML, et al. Nitrogen-Doped Mesoporous Carbon Promoted Chemical Adsorption of Sulfur and Fabrication of High-Areal-Capacity Sulfur Cathode with Exceptional Cycling Stability for Lithium-Sulfur Batteries. *Adv Funct Mater.* 2014;24:1243–1250.
73 Li L, Zhou G, Yin L, et al. Stabilizing Sulfur Cathodes Using Nitrogen-Doped Graphene as a Chemical Immobilizer for Li[Formula presented]S Batteries. *Carbon.* 2016;108:120–126.
74] Xiang M, Yang L, Zheng Y, et al. A Freestanding and Flexible Nitrogen-Doped Carbon Foam/Sulfur Cathode Composited with Reduced Graphene Oxide for High Sulfur Loading Lithium-Sulfur Batteries. *J Mater Chem A.* 2017;5:18020–18028.
75 Tan J, Li D, Liu Y, et al. A Self-Supported 3D Aerogel Network Lithium-Sulfur Battery Cathode: Sulfur Spheres Wrapped with Phosphorus Doped Graphene and Bridged with Carbon Nanofibers. *J Mater Chem A.* 2020;8:7980–7990.
76 Liu F, Wang N, Shi C, et al. Phosphorus Doping of 3D Structural MoS2 to Promote Catalytic Activity for Lithium-Sulfur Batteries. *Chem Eng J.* 2022;431:133923.
77 Xie Y, Meng Z, Cai T, et al. Effect of Boron-Doping on the Graphene Aerogel Used as Cathode for the Lithium-Sulfur Battery. *ACS Appl Mater Interfaces.* 2015;7:25202–25210.
78 Yang CP, Yin YX, Ye H, et al. Insight into the Effect of Boron Doping on Sulfur/Carbon Cathode in Lithium-Sulfur Batteries. *ACS Appl Mater Interfaces.* 2014;6:8789–8795.
79 Qiang Z, Chen YM, Xia Y, et al. Ultra-Long Cycle Life, Low-Cost Room Temperature Sodium-Sulfur Batteries Enabled by Highly Doped (N,S) Nanoporous Carbons. *Nano Energy.* 2017;32:59–66.
80 Hao Y, Li X, Sun X, et al. Nitrogen-Doped Graphene Nanosheets/S Composites as Cathode in Room-Temperature Sodium-Sulfur Batteries. *ChemistrySelect.* 2017;2:9425–9432.
81 Andritsos EI, Lekakou C, Cai Q. Single-Atom Catalysts as Promising Cathode Materials for Lithium-Sulfur Batteries. *J Phys Chem C.* 2021;125:18108–18118.
82 Zhuang Z, Kang Q, Wang D, et al. Single-Atom Catalysis Enables Long-Life, High-Energy Lithium-Sulfur Batteries. *Nano Res.* 2020;13:1856–1866.
83 Xiao R, Chen K, Zhang X, et al. Single-Atom Catalysts for Metal-Sulfur Batteries: Current Progress and Future Perspectives. *J Energy Chem.* 2021;54:452–466.
84 Wang F, Li J, Zhao J, et al. Single-Atom Electrocatalysts for Lithium Sulfur Batteries: Progress, Opportunities, and Challenges. *ACS Mater Lett.* 2020;2:1450–1463.

85 Zhang BW, Sheng T, Wang YX, et al. Long-Life Room-Temperature Sodium–Sulfur Batteries by Virtue of Transition-Metal-Nanocluster–Sulfur Interactions. *Angew Chemie Int Ed*. 2019;58:1484–1488.
86 Zhang BW, Sheng T, Liu YD, et al. Atomic Cobalt as an Efficient Electrocatalyst in Sulfur Cathodes for Superior Room-Temperature Sodium-Sulfur Batteries. *Nat Commun*. 2018;9:4082.
87 Guo B, Du W, Yang T, et al. Nickel Hollow Spheres Concatenated by Nitrogen-Doped Carbon Fibers for Enhancing Electrochemical Kinetics of Sodium–Sulfur Batteries. *Adv Sci*. 2020;7:1902617.
88 Zhu J, Abdelkader A, Demko D, et al. Electrocatalytic Assisted Performance Enhancement for the Na-S Battery in Nitrogen-Doped Carbon Nanospheres Loaded with Fe. *Molecules*. 2020;25:1585.
89 Ma Q, Du G, Guo B, et al. Carbon-Wrapped Cobalt Nanoparticles on Graphene Aerogel for Solid-State Room-Temperature Sodium-Sulfur Batteries. *Chem Eng J*. 2020;388:124210.
90 Wang N, Wang Y, Bai Z, et al. High-Performance Room-Temperature Sodium–Sulfur Battery Enabled by Electrocatalytic Sodium Polysulfides Full Conversion. *Energy Environ Sci*. 2020;13:562–570.
91 Du W, Shen K, Qi Y, et al. Efficient Catalytic Conversion of Polysulfides by Biomimetic Design of "Branch-Leaf" Electrode for High-Energy Sodium–Sulfur Batteries. *Nano-Micro Lett*. 2021;13:50.
92 Yan Z, Liang Y, Hua W, et al. Multiregion Janus-Featured Cobalt Phosphide-Cobalt Composite for Highly Reversible Room-Temperature Sodium-Sulfur Batteries. *ACS Nano*. 2020;14:10284–10293.
93 Xiao F, Wang H, Yao T, et al. MOF-Derived CoS_2/N-Doped Carbon Composite to Induce Short-Chain Sulfur Molecule Generation for Enhanced Sodium-Sulfur Battery Performance. *ACS Appl Mater Interfaces*. 2021;13:18010–18020.
94 Aslam MK, Seymour ID, Katyal N, et al. Metal Chalcogenide Hollow Polar Bipyramid Prisms as Efficient Sulfur Hosts for Na-S Batteries. *Nat Commun*. 2020;11:1–11.
95 Zheng Y, Yi Y, Fan M, et al. A High-Entropy Metal Oxide as Chemical Anchor of Polysulfide for Lithium-Sulfur Batteries. *Energy Storage Mater*. 2019;23:678–683.
96 Marangon V, Scaduti E, Vinci VF, et al. Scalable Composites Benefiting from Transition-Metal Oxides as Cathode Materials for Efficient Lithium-Sulfur Batteries. *ChemElectroChem*. 2022;9:e202200374.
97 Liu X, Huang JQ, Zhang Q, et al. Nanostructured Metal Oxides and Sulfides for Lithium–Sulfur Batteries. *Adv Mater*. 2017;29:1601759.
98 Liang X, Kwok CY, Lodi-Marzano F, et al. Tuning Transition Metal Oxide-Sulfur Interactions for Long Life Lithium Sulfur Batteries: The "goldilocks" Principle. *Adv Energy Mater*. 2016;6:1–9.
99 Wang J, Li G, Luo D, et al. Engineering the Conductive Network of Metal Oxide-Based Sulfur Cathode toward Efficient and Longevous Lithium–Sulfur Batteries. *Adv Energy Mater*. 2020;10:1–11.
100 Ma D, Li Y, Yang J, et al. New Strategy for Polysulfide Protection Based on Atomic Layer Deposition of TiO_2 onto Ferroelectric-Encapsulated Cathode: Toward Ultrastable Free-Standing Room Temperature Sodium–Sulfur Batteries. *Adv Funct Mater*. 2018;28:14–16.
101 Ye X, Ruan J, Pang Y, et al. Enabling a Stable Room-Temperature Sodium-Sulfur Battery Cathode by Building Heterostructures in Multichannel Carbon Fibers. *ACS Nano*. 2021;15:5639–5648.
102 Chen L, Li X, Xu Y. Recent Advances of Polar Transition-Metal Sulfides Host Materials for Advanced Lithium-Sulfur Batteries. *Funct Mater Lett*. 2018;11:1840010.

103 Zuo JH, Gong YJ. Applications of Transition-Metal Sulfides in the Cathodes of Lithium–Sulfur Batteries. *Tungsten.* 2020;2:134–146.
104 Ye H, Ma L, Zhou Y, et al. Amorphous MoS_3 as the Sulfur-Equivalent Cathode Material for Room-Temperature Li–S and Na–S Batteries. *Proc Natl Acad Sci USA.* 2017;114:13091–13096.
105 Meyerson ML, Papa PE, Weeks JA, et al. Sulfur-Rich Molybdenum Sulfide as a Cathode Material for Room Temperature Sodium-Sulfur Batteries. *ACS Appl Energy Mater.* 2020;3:6121–6126.
106 Yan Z, Xiao J, Lai W, et al. Nickel Sulfide Nanocrystals on Nitrogen-Doped Porous Carbon Nanotubes with High-Efficiency Electrocatalysis for Room-Temperature Sodium-Sulfur Batteries. *Nat Commun.* 2019;10:1–8.
107 Liu H, Pei W, Lai WH, et al. Electrocatalyzing S Cathodes via Multisulfiphilic Sites for Superior Room-Temperature Sodium-Sulfur Batteries. *ACS Nano.* 2020;14:7259–7268.
108 Rauh RD, Abraham KM, Pearson GF, et al. A Lithium/Dissolved Sulfur Battery with an Organic Electrolyte. *J Electrochem Soc.* 1979;126:523–527.
109 Yu X, Manthiram A. Na_2S-Carbon Nanotube Fabric Electrodes for Room-Temperature Sodium-Sulfur Batteries. *Chem Eur J.* 2015;21:4233–4237.
110 Yu X, Manthiram A. Performance Enhancement and Mechanistic Studies of Room-Temperature Sodium-Sulfur Batteries with a Carbon-Coated Functional Nafion Separator and a Na_2S/Activated Carbon Nanofiber Cathode. *Chem Mater.* 2016;28:896–905.
111 Wang C, Wang H, Hu X, et al. Frogspawn-Coral-Like Hollow Sodium Sulfide Nanostructured Cathode for High-Rate Performance Sodium–Sulfur Batteries. *Adv Energy Mater.* 2019;9:1–9.
112 Bloi LM, Pampel J, Dörfler S, et al. Sodium Sulfide Cathodes Superseding Hard Carbon Pre-sodiation for the Production and Operation of Sodium–Sulfur Batteries at Room Temperature. *Adv Energy Mater.* 2020;10:1903245.
113 Kumar A, Ghosh A, Roy A, et al. High-Energy Density Room Temperature Sodium-Sulfur Battery Enabled by Sodium Polysulfide Catholyte and Carbon Cloth Current Collector Decorated with MnO_2 Nanoarrays. *Energy Storage Mater.* 2019;20:196–202.
114 Bharti VK, Pathak AD, Anjan A, et al. Covalently Confined Sulfur Composite with Carbonized Bacterial Cellulose as an Efficient Cathode Matrix for High-Performance Potassium-Sulfur Batteries. *ACS Sustain Chem Eng.* 2022;10:16634–16646.
115 Ahmed MS, Lee S, Agostini M, et al. Multiscale Understanding of Covalently Fixed Sulfur–Polyacrylonitrile Composite as Advanced Cathode for Metal–Sulfur Batteries. *Adv Sci.* 2021;8:1–34.
116 Wu T, Jing M, Yang L, et al. Controllable Chain-Length for Covalent Sulfur–Carbon Materials Enabling Stable and High-Capacity Sodium Storage. *Adv Energy Mater.* 2019;9:1–11.
117 Yan J, Li W, Wang R, et al. An in Situ Prepared Covalent Sulfur-Carbon Composite Electrode for High-Performance Room-Temperature Sodium-Sulfur Batteries. *ACS Energy Lett.* 2020;5:1307–1315.
118 Chen K, Li HJW, Xu Y, et al. Untying Thioether Bond Structures Enabled by "Voltage-Scissors" for Stable Room Temperature Sodium-Sulfur Batteries. *Nanoscale.* 2019;11:5967–5973.
119 Hu H, Cheng H, Liu Z, et al. In Situ Polymerized PAN-Assisted S/C Nanosphere with Enhanced High-Power Performance as Cathode for Lithium/Sulfur Batteries. *Nano Lett.* 2015;15:5116–5123.
120 Ye J, He F, Nie J, et al. Sulfur/Carbon Nanocomposite-Filled Polyacrylonitrile Nanofibers as a Long Life and High Capacity Cathode for Lithium-Sulfur Batteries. *J Mater Chem A.* 2015;3:7406–7412.

121 Li Z, Zhang J, Lu Y, et al. A Pyrolyzed Polyacrylonitrile/Selenium Disulfide Composite Cathode with Remarkable Lithium and Sodium Storage Performances. *Sci Adv.* 2018;4:1–11.
122 Ghosh A, Shukla S, Monisha M, et al. Sulfur Copolymer: A New Cathode Structure for Room-Temperature Sodium-Sulfur Batteries. *ACS Energy Lett.* 2017;2:2478–2485.
123 Li S, Zeng Z, Yang J, et al. High Performance Room Temperature Sodium–Sulfur Battery by Eutectic Acceleration in Tellurium-Doped Sulfurized Polyacrylonitrile. *ACS Appl Energy Mater.* 2019;2:2956–2964.
124 Ghosh A, Kumar A, Roy A, et al. Three-Dimensionally Reinforced Freestanding Cathode for High-Energy Room-Temperature Sodium-Sulfur Batteries. *ACS Appl Mater Interfaces.* 2019;11:14101–14109.
125 Huang Z, Song B, Zhang H, et al. High-Capacity and Stable Sodium-Sulfur Battery Enabled by Confined Electrocatalytic Polysulfides Full Conversion. *Adv Funct Mater.* 2021;31:2100666.
126 Zhang SS. Liquid Electrolyte Lithium/Sulfur Battery: Fundamental Chemistry, Problems, and Solutions. *J Power Sources.* 2013;231:153–162.
127 Mu P, Dong T, Jiang H, et al. Crucial Challenges and Recent Optimization Progress of Metal-Sulfur Battery Electrolytes. *Energy Fuels.* 2021;35:1966–1988.
128 Pan Y, Li S, Yin M, et al. Electrolyte Evolution Propelling the Development of Nonlithium Metal–Sulfur Batteries. *Energy Technol.* 2019;7:1900164.
129 Soni CB, Sungjemmenla, Vineeth SK, et al. Challenges in Regulating Interfacial-Chemistry of the Sodium-Metal Anode for Room-Temperature Sodium-Sulfur Batteries. *Energy Storage.* 2022;4:e264.
130 Vineeth SK, Soni CB, Sun Y, et al. Implications of Na-Ion Solvation on Na Anode–Electrolyte Interphase. *Trends Chem.* 2022;4:48–59.
131 Eng AYS, Kumar V, Zhang Y, et al. Room-Temperature Sodium–Sulfur Batteries and Beyond: Realizing Practical High Energy Systems through Anode, Cathode, and Electrolyte Engineering. *Adv Energy Mater.* 2021;11:2003493.
132 Cha E, Patel M, Bhoyate S, et al. Nanoengineering to Achieve High Efficiency Practical Lithium-Sulfur Batteries. *Nanoscale Horizons.* 2020;5:808–831.
133 Li M, Zhang Y, Bai Z, et al. A Lithium–Sulfur Battery using a 2D Current Collector Architecture with a Large-Sized Sulfur Host Operated under High Areal Loading and Low E/S Ratio. *Adv Mater.* 2018;30:1–9.
134 Luo C, Hu E, Gaskell KJ, et al. A Chemically Stabilized Sulfur Cathode for Lean Electrolyte Lithium Sulfur Batteries. *Proc Natl Acad Sci USA.* 2020;117:14712–14720.
135 Zhang SS. Improved Cyclability of Liquid Electrolyte Lithium/Sulfur Batteries by Optimizing Electrolyte/Sulfur Ratio. *Energies.* 2012;5:5190–5197.
136 Hagen M, Hanselmann D, Ahlbrecht K, et al. Lithium-Sulfur Cells: The Gap between the State-of-the-Art and the Requirements for High Energy Battery Cells. *Adv Energy Mater.* 2015;5:1401986.
137 Bhargav A, He J, Gupta A, et al. Lithium-Sulfur Batteries: Attaining the Critical Metrics. *Joule.* 2020;4:285–291.
138 Eroglu D, Zavadil KR, Gallagher KG. Critical Link between Materials Chemistry and Cell-Level Design for High Energy Density and Low Cost Lithium-Sulfur Transportation Battery. *J Electrochem Soc.* 2015;162:A982–A990.
139 Betz J, Bieker G, Meister P, et al. Theoretical versus Practical Energy: A Plea for More Transparency in the Energy Calculation of Different Rechargeable Battery Systems. *Adv Energy Mater.* 2019;9:1–18.
140 Guo W, Fu Y. A Perspective on Energy Densities of Rechargeable Li-S Batteries and Alternative Sulfur-Based Cathode Materials. *Energy Environ Mater.* 2018;1:20–27.
141 Bilal HM, Eroglu D. Assessment of Li-S Battery Performance as a Function of Electrolyte-to-Sulfur Ratio. *J Electrochem Soc.* 2021;168:030502.

142 Pope MA, Aksay IA. Structural Design of Cathodes for Li-S Batteries. *Adv Energy Mater.* 2015;5:1500124.
143 Liu Y, Liu S, Li G-R, et al. High Volumetric Energy Density Sulfur Cathode with Heavy and Catalytic Metal Oxide Host for Lithium–Sulfur Battery. *Adv Sci.* 2020;7:1903693.
144 Han P, Chung SH, Manthiram A. Designing a High-Loading Sulfur Cathode with a Mixed Ionic-Electronic Conducting Polymer for Electrochemically Stable Lithium-Sulfur Batteries. *Energy Storage Mater.* 2019;17:317–324.
145 Kong L, Jin Q, Zhang XT, et al. Towards Full Demonstration of High Areal Loading Sulfur Cathode in Lithium–Sulfur Batteries. *J Energy Chem.* 2019;39:17–22.
146 Luo L, Manthiram A. Rational Design of High-Loading Sulfur Cathodes with a Poached-Egg-Shaped Architecture for Long-Cycle Lithium-Sulfur Batteries. *ACS Energy Lett.* 2017;2:2205–2211.
147 Kumar D, Kanchan DK, Kumar S, et al. Recent Trends on Tailoring Cathodes for Room-Temperature Na-S Batteries. *Mater Sci Energy Technol.* 2019;2:117–129.
148 Lu LQ, Lu LJ, Wang Y. Sulfur Film-Coated Reduced Graphene Oxide Composite for Lithium-Sulfur Batteries. *J Mater Chem A.* 2013;1:9173–9181.
149 Ogoke O, Wu G, Wang X, et al. Effective Strategies for Stabilizing Sulfur for Advanced Lithium-Sulfur Batteries. *J Mater Chem A.* 2017;5:448–469.
150 Setiawan B, Iasha V, Hikmah U. Lithium-Sulfur Battery: The Review of Cathode Composite Fabrication Method. *Int J Sci Technol Res.* 2020;9:737–741.
151 Xu J, Shui J, Wang J, et al. Sulfur-Graphene Nanostructured Cathodes via Ball-Milling for High. *ACS Nano.* 2014;8:10920–10930.
152 Kang HS, Sun YK. Freestanding Bilayer Carbon-Sulfur Cathode with Function of Entrapping Polysulfide for High Performance Li-S Batteries. *Adv Funct Mater.* 2016;26:1225–1232.
153 Zhang K, Zhao Q, Tao Z, et al. Composite of Sulfur Impregnated in Porous Hollow Carbon Spheres as the Cathode of Li-S Batteries with High Performance. *Nano Res.* 2013;6:38–46.
154 Tian C, Wu J, Ma Z, et al. A Melt-Diffusion Strategy for Tunable Sulfur Loading on CC@MoS_2 for Lithium–Sulfur Batteries. *Energy Rep.* 2020;6:172–180.
155 Lu Y, Li X, Liang J, et al. A Simple Melting-Diffusing-Reacting Strategy to Fabricate S/NiS_2-C for Lithium-Sulfur Batteries. *Nanoscale.* 2016;8:17616–17622.
156 Chen S-R, Zhai Y-P, Xu G-L, et al. Ordered Mesoporous Carbon/Sulfur Nanocomposite of High Performances as Cathode for Lithium–Sulfur Battery. *Electrochim Acta.* 2011;56:9549–9555.
157 Choudhury S, Zeiger M, Massuti-Ballester P, et al. Carbon Onion-Sulfur Hybrid Cathodes for Lithium-Sulfur Batteries. *Sustain Energy Fuels.* 2017;1:84–94.
158 Xia Y, Zhong H, Fang R, et al. Biomass Derived $Ni(OH)_2$@Porous Carbon/Sulfur Composites Synthesized by a Novel Sulfur Impregnation Strategy Based on Supercritical CO_2 Technology for Advanced Li-S Batteries. *J Power Sources.* 2018;378:73–80.
159 Li M, Carter R, Douglas A, et al. Sulfur Vapor-Infiltrated 3D Carbon Nanotube Foam for Binder-Free High Areal Capacity Lithium-Sulfur Battery Composite Cathodes. *ACS Nano.* 2017;11:4877–4884.
160 Carter R, Davis B, Oakes L, et al. A High Areal Capacity Lithium-Sulfur Battery Cathode Prepared by Site-Selective Vapor Infiltration of Hierarchical Carbon Nanotube Arrays. *Nanoscale.* 2017;9:15018–15026.
161 Zhou K, Fan XJ, Wei XF, et al. The Strategies of Advanced Cathode Composites for Lithium-Sulfur Batteries. *Sci China Technol Sci.* 2017;60:175–185.
162 Ye Y, Wu F, Xu S, et al. Designing Realizable and Scalable Techniques for Practical Lithium Sulfur Batteries: A Perspective. *J Phys Chem Lett.* 2018;9:1398–1414.

163 Fan X, Zhang Y, Li J, et al. A General Dissolution-Recrystallization Strategy to Achieve Sulfur-Encapsulated Carbon for an Advanced Lithium-Sulfur Battery. *J Mater Chem A*. 2018;6:11664–11669.

164 Fan X, Chen F, Zhang Y, et al. Constructing a LiPAA Interface Layer: A New Strategy to Suppress Polysulfide Migration and Facilitate Li+ Transport for High-Performance Flexible Li-S Batteries. *Nanotechnology*. 2020;31:095401.

165 Zha C, Wu D, Zhang T, et al. A Facile and Effective Sulfur Loading Method: Direct Drop of Liquid Li_2S_8 on Carbon Coated TiO_2 Nanowire Arrays as Cathode towards Commercializing Lithium-Sulfur Battery. *Energy Storage Mater*. 2019; 17:118–125.

166 Gadhave RV, Vineeth SK. Synthesis and Characterization of Starch Stabilized Polyvinyl Acetate-Acrylic Acid Copolymer-Based Wood Adhesive. *Polym Bull*. 2022. https://doi.org/10.1007/s00289-022-04558-8.

167 Vineeth SK, Gadhave RV. Corn Starch Blended Polyvinyl Alcohol Adhesive Chemically Modified by Crosslinking and Its Applicability as Polyvinyl Acetate Wood Adhesive. *Polym Bull*. 2023. https://doi.org/10.1007/s00289-023-04746-0.

168 Vineeth SK, Gadhave RV, Gadekar PT. Glyoxal Cross-Linked Polyvinyl Alcohol-Microcrystalline Cellulose Blend as a Wood Adhesive with Enhanced Mechanical, Thermal and Performance Properties. *Mater Int*. 2020;2:0277–0285.

169 Chen YM, Liang W, Li S, et al. A Nitrogen Doped Carbonized Metal-Organic Framework for High Stability Room Temperature Sodium-Sulfur Batteries. *J Mater Chem A*. 2016;4:12471–12478.

170 Xiao F, Yang X, Wang H, et al. Covalent Encapsulation of Sulfur in a MOF-Derived S, N-Doped Porous Carbon Host Realized via the Vapor-Infiltration Method Results in Enhanced Sodium–Sulfur Battery Performance. *Adv Energy Mater*. 2020;10:1–10.

171 Lei Y, Wu C, Lu X, et al. Streamline Sulfur Redox Reactions to Achieve Efficient Room-Temperature Sodium–Sulfur Batteries. *Angew Chemie Int Ed*. 2022;61:e202200384.

172 Liu Y, Ma S, Rosebrock M, et al. Tungsten Nanoparticles Accelerate Polysulfides Conversion: A Viable Route toward Stable Room-Temperature Sodium–Sulfur Batteries. *Adv Sci*. 2022;9:1–8.

173 Wang H, Qi Y, Xiao F, et al. Tessellated N-Doped Carbon/$CoSe_2$ as Trap-Catalyst Sulfur Hosts for Room-Temperature Sodium-Sulfur Batteries. *Inorg Chem Front*. 2022;9:1743–1751.

174 Wu J, Yu Z, Yao Y, et al. Bifunctional Catalyst for Liquid–Solid Redox Conversion in Room-Temperature Sodium–Sulfur Batteries. *Small Struct*. 2022;3:2200020.

175 Kim I, Kim CH, Choi SH, et al. A Singular Flexible Cathode for Room Temperature Sodium/Sulfur Battery. *J Power Sources*. 2016;307:31–37.

176 Eng AYS, Nguyen DT, Kumar V, et al. Tailoring Binder-Cathode Interactions for Long-Life Room-Temperature Sodium-Sulfur Batteries. *J Mater Chem A*. 2020;8:22983–22997.

177 Hu P, Xiao F, Wu Y, et al. Covalent Encapsulation of Sulfur in a Graphene/N-Doped Carbon Host for Enhanced Sodium-Sulfur Batteries. *Chem Eng J*. 2022;443:136257.

178 Chen J, Zhang H, Yang H, et al. Towards Practical Li–S Battery with Dense and Flexible Electrode Containing Lean Electrolyte. *Energy Storage Mater*. 2020;27:307–315.

179 Yang H, Chen J, Yang J, et al. Dense and High Loading Sulfurized Pyrolyzed Poly (Acrylonitrile)(S@pPAN) Cathode for Rechargeable Lithium Batteries. *Energy Storage Mater*. 2020;31:187–194.

180 Qie L, Manthiram A. High-Energy-Density Lithium-Sulfur Batteries Based on Blade-Cast Pure Sulfur Electrodes. *ACS Energy Lett*. 2016;1:46–51.

181 Verdier N, Foran G, Lepage D, et al. Challenges in Solvent-Free Methods for Manufacturing Electrodes and Electrolytes for Lithium-Based Batteries. *Polymers (Basel)*. 2021;13:1–26.

182 Zhu C, Fu Y, Yu Y. Designed Nanoarchitectures by Electrostatic Spray Deposition for Energy Storage. *Adv Mater.* 2019;31:1–25.
183 Rabiei Baboukani A, Khakpour I, Adelowo E, et al. High-Performance Red Phosphorus-Sulfurized Polyacrylonitrile Composite by Electrostatic Spray Deposition for Lithium-Ion Batteries. *Electrochim Acta.* 2020;345:136227.
184 Li S-R, Yesibolati N, Qiao Y, et al. Electrostatic Spray Deposition of Porous $Fe_2V_4O_{13}$ Films as Electrodes for Li-Ion Batteries. *J Alloys Compd.* 2012;520:77–82.
185 Ding B, Yuan C, Shen L, et al. Encapsulating Sulfur into Hierarchically Ordered Porous Carbon as a High-Performance Cathode for Lithium-Sulfur Batteries. *Chem Eur J.* 2013;19:1013–1019.
186 Tian L, Wang M, Xiong L, et al. Preparation and Performance of p(OPal-MMA)/PVDF Blend Polymer Membrane via Phase-Inversion Process for Lithium-Ion Batteries. *J Electroanal Chem.* 2019;839:264–273.
187 Choudhury S, Fischer D, Formanek P, et al. Porous Carbon Prepared from Polyacrylonitrile for Lithium-Sulfur Battery Cathodes Using Phase Inversion Technique. *Polymer (Guildf).* 2018;151:171–178.
188 Pu W, He X, Wang L, et al. Preparation of PVDF-HFP Microporous Membrane for Li-Ion Batteries by Phase Inversion. *J Memb Sci.* 2006;272:11–14.
189 Choudhury S, Fischer D, Formanek P, et al. Phase Inversion Strategy to Fabricate Porous Carbon for Li-S Batteries via Block Copolymer Self-Assembly. *Adv Mater Interfaces.* 2018;5:1–11.
190 Yang X, Chen Y, Wang M, et al. Phase Inversion: A Universal Method to Create High-Performance Porous Electrodes for Nanoparticle-Based Energy Storage Devices. *Adv Funct Mater.* 2016;26:8427–8434.
191 Gao X, Sun Q, Yang X, et al. Toward a Remarkable Li-S Battery via 3D Printing. *Nano Energy.* 2019;56:595–603.
192 Chen C, Jiang J, He W, et al. 3D Printed High-Loading Lithium-Sulfur Battery Toward Wearable Energy Storage. *Adv Funct Mater.* 2020;30:1–7.
193 Aurbach D. Review of Selected Electrode–Solution Interactions which Determine the Performance of Li and Li Ion Batteries. *J Power Sources.* 2000;89:206–218.
194 Rehman S, Pope M, Tao S, et al. Evaluating the Effectiveness of In Situ Characterization Techniques in Overcoming Mechanistic Limitations in Lithium-Sulfur Batteries. *Energy Environ Sci.* 2022;15:1423–1460.
195 Liu X, Tong Y, Wu Y, et al. In-Depth Mechanism Understanding for Potassium-Ion Batteries by Electroanalytical Methods and Advanced In Situ Characterization Techniques. *Small Methods.* 2021;5:1–22.
196 Li H, Guo S, Zhou H. In-Situ/Operando Characterization Techniques in Lithium-Ion Batteries and Beyond. *J Energy Chem.* 2021;59:191–211.
197 Yang F, Feng X, Liu YS, et al. In Situ/Operando (Soft) X-ray Spectroscopy Study of Beyond Lithium-ion Batteries. *Energy Environ Mater.* 2021;4:139–157.
198 Yan Y, Cheng C, Zhang L, et al. Deciphering the Reaction Mechanism of Lithium–Sulfur Batteries by In Situ/Operando Synchrotron-Based Characterization Techniques. *Adv Energy Mater.* 2019;9:1–14.
199 Conder J, Bouchet R, Trabesinger S, et al. Direct Observation of Lithium Polysulfides in Lithium-Sulfur Batteries Using Operando X-Ray Diffraction. *Nat Energy.* 2017;2:17069.
200 Lang S, Yu SH, Feng X, et al. Understanding the Lithium–Sulfur Battery Redox Reactions via Operando Confocal Raman Microscopy. *Nat Commun.* 2022;13:1–11.
201 Jia L, Wang J, Ren S, et al. Unraveling Shuttle Effect and Suppression Strategy in Lithium/Sulfur Cells by In Situ/Operando X-ray Absorption Spectroscopic Characterization. *Energy Environ Mater.* 2021;4:222–228.
202 Mahankali K, Thangavel NK, Reddy Arava LM. In Situ Electrochemical Mapping of Lithium-Sulfur Battery Interfaces Using AFM-SECM. *Nano Lett.* 2019;19:5229–5236.

203 Parke CD, Teo L, Schwartz DT, et al. Progress on Continuum Modeling of Lithium-Sulfur Batteries. *Sustain Energy Fuels*. 2021;5:5946–5966.
204 Lin Y, Ticey J, Oleshko V, et al. Carbon-Nanotube-Encapsulated-Sulfur Cathodes for Lithium-Sulfur Batteries: Integrated Computational Design and Experimental Validation. *Nano Lett*. 2022;22:441–447.
205 Zhang L, Zhang W, Ma X, et al. Computational Screening of Functionalized MXenes to Catalyze the Solid and Non-Solid Conversion Reactions in Cathodes of Lithium-Sulfur Batteries. *Phys Chem Chem Phys*. 2022;24:8913–8922.
206 Abraham AM, Boteju T, Ponnurangam S, et al. A Global Design Principle for Polysulfide Electrocatalysis in Lithium–Sulfur Batteries—A Computational Perspective. *Batter Energy*. 2022;1:20220003.
207 Grabe S, Baboo JP, Tennison S, et al. Sulfur Infiltration and Allotrope Formation in Porous Cathode Hosts for Lithium-Sulfur Batteries. *AIChE J*. 2022;68:1–12.

4 Electrolytes for room-temperature sodium-sulfur batteries
A holistic approach to understand solvation

SK Vineeth, Sungjemmenla, Yusuke Yamauchi, and Vipin Kumar

4.1 BASIC PROPERTIES OF THE ELECTROLYTES FOR ALKALI METAL BATTERIES

Renewable sources such as solar, wind, tidal, and geothermal contribute less to the carbon footprint; however, the bottleneck issue lies in their intermittence nature. Hence, energy storage devices have successfully gained ground to ensure a constant supply of power, irrespective of the availability of the Sun or wind [1]. Amongst various storage devices, rechargeable electrochemical batteries are a prime solution for efficient and mobile storage [2–5]. Alkali metal batteries employ alkali metal anode, for instance, lithium (Li), sodium (Na), and potassium (K) [6]. The electrolyte requirements for the alkali metal anodes are primarily different from the conventional graphite-based anode (i.e., LiC_6). Developing electrolyte formulation for alkali-metal batteries involves a reasonable selection of salt and solvent systems. In metal-sulfur batteries, for instance, the metal anode, e.g., lithium or sodium, reacts spontaneously with the organic solvents. A reasonable consideration, which may include the concept of donor and acceptor number, anionic size of the electrolyte salt, ionic conductivity, relative molecular energy levels of the solvent, and chemical compatibility of the salt and solvent, should be made while designing the electrolyte systems. If it remains ignored, it may impart deterioration in cell performance and lead to the decomposition of electrolytes or, eventually, cell failure.

As one of the consideration parameters, the suitability of the electrolyte system can be determined by the donor number (DN) or donocity of the solvent, which relates to the nucleophilic nature [7]. In other words, how easily solvent molecules solvate the cations of the salt is determined by the DN. In addition, Gutmann postulated another parameter called acceptor number (AN), which describes electrophilic behavior [8]. It can also be through the ease of solvating the anions of the salt. Often ignored, these parameters assist in unveiling the ion-solvation behavior of electrolytes. For metal-sulfur batteries, it is identified that a solvent with a higher DN number endorses redox reactions by altering the kinetic pathways and is observed to stabilize polysulfide anions [9]. In particular, for lithium-sulfur batteries, dimethylacetamide (DMA) is ideal for the full utilization of sulfur with lithium disulfide as the discharge product [10]. However, alkali metals react highly toward the high DN solvents, which impedes their widespread usage in metal anode batteries [11]. As a result, solvents with high DN show poor cycling performance in the case of metal-sulfur batteries, and therefore, the practical usage of high DN solvents is limited. Alteration of Gutmann AN has shown to be an effective strategy in enhancing the stability of discharge products formed during discharge reactions [7]. Solvents with a high AN exhibit good solubility of the discharge products, effectively achieving long-term stability and high energy density [12]. For example, 2–3 M of sodium trifluoromethane sulfonate (NaOTf) can be solvated in diglyme, while only 1–2 M of NaOtf can be solvated in tetraglyme. This behavior is linked with the DN and AC of the solvents.

Since a definite electronegativity difference exists between solvent molecules and metal cations, a short-range interaction develops between them. For insightful knowledge of the ion transportation mechanism in electrolytes, parameters affecting ionic conduction need to be divulged. Ionic conductivity signifies the movement of mobile ions under the influence of an applied electric field. It can be equated as the product of ion concentration, the charge of ions, and the mobility of ions. Although there is a difference in the ionic radii of Li^+ and Na^+, a few similarities can be seen in the interactions between alkali-ions (Na^+ and Li^+) and electrolyte molecules [13,14]. However, the solvation structures of Li^+ and Na^+ are entirely different. Li^+ orients in a tetrahedral geometry, while the solvation structure of Na^+ is not well ordered [15]. Thus a generalization on the performance of electrolyte systems working well for lithium-based batteries cannot be assured in the case of sodium metal batteries (SMBs) [16–18]. The choice of selection of electrolyte solvent also regulates the performance of SMBs.

Ideally, the requirement of a solvent includes low viscosity, high dielectric constant value, and optimum values of Lewis's acidity/basicity, thereby favoring ionic conductivity. Electrolyte solvents with strong Lewis's basicity promote the solvation of sodium salts. This phenomenon is due to the coordination between the empty orbitals of Na^+ and lone pairs of electrolyte solvent molecules, which enhances the salt solubility in the solvent. If the interactions are very strong, the de-solvation process could be affected negatively, thereby leading to co-intercalation between Na^+ and solvent molecules [19]. Another parameter that signifies the applicability of an electrolyte solvent is its electrochemical stability. To quantify the electrochemical stability of an electrolyte, a terminology known as electrochemical stability window (ESW) is defined. Concerning the electrolyte, ESW is the potential range within which the electrolyte material is neither reduced nor oxidized. Energy levels of solvent molecules, such as highest occupied molecular orbital (HOMO) and lowest unoccupied molecular orbital (LUMO), play a significant role in determining electrolyte reduction. This is relative to the redox reactions of the electrode material [17]. When the electron energies are higher than the LUMO level, solvent molecules

TABLE 4.1
Various properties of electrolyte solvents and their respective LUMO and HOMO values

Electrolyte solvent	Viscosity (cP) at 25°C	Dielectric constant at 25°C	Flash point (°C)	LUMO	HOMO
Propylene carbonate (PC)	2.53	64.92	~140	−0.0149	−0.2547
Ethylene carbonate (EC)	1.90	89.78	~165	−0.0177	−0.2585
Dimethyl carbonate (DMC)	0.59	3.107	~20	−0.0091	−0.2488
Ethyl methyl carbonate (EMC)	0.65	2.958	~30	−0.0062	−0.2457
Diethyl carbonate (DEC)	0.75	2.805	~40	−0.0036	−0.2426

are reduced, and if the value falls below the HOMO value the solvent gets oxidized. To unveil the complexity and to make the concept of ESW simplified, Peljo et al. revisited the definition of ESW as the difference of oxidation and reduction potential of solvent molecules [20]. Other desired properties of electrolyte solvent include safety aspects such as high flash point and boiling point, low melting point, and lower toxicity. Tables 4.1 and 4.2 summarize various properties of electrolyte solvents and salts used in SMBs.

4.1.1 Evolution of the Room-Temperature Sodium-Sulfur Battery – Transition from a Solid-Electrolyte to the Liquid Electrolyte

The chemistry of Na-S was invented in the 1960s; the system was called a high-temperature Na-S (HT Na-S) battery in the 1960s. It was operated at temperatures ~300°C, containing molten Na as the anode, beta-alumina as the electrolyte, and molten sulfur-impregnated cathode [21]. At room temperature, the ionic conductivity of BASE barely supports the electrochemical reactions. A minimum operating temperature of 250°C–300°C is required to ensure the reversible function of Na-S chemistry. The HT Na-S battery with a ceramic electrolyte consisting of β''-Al_2O_3 solid electrolyte (BASE) was developed in 1966 by Kummer et al. at Ford Motor Company [22,23]. Due to the high energy density of HT Na-S, the battery chemistry was considered prominent for mobility applications. However, the high operating temperature posed serious safety challenges, and the battery technology was never deployed for mobility applications. In 1976, Goodenough et al. developed an inorganic solid electrolyte (ISE) for sodium ion (Na^+) conduction called NAtrium SuperIonic CONductors (NASICON), which performed equivalent to BASE [24]. However, the operating temperature could not be lowered significantly. Despite its high-temperature operation, NGK Insulators, Ltd

TABLE 4.2
Properties of sodium salts reported for sodium metal batteries with chemical structure and HOMO level values

Sodium salt	Chemical structure	HOMO level
Sodium perchlorate ($NaClO_4$)		−7.89
Sodium hexafluorophosphate ($NaPF_6$)		−11.67
Sodium trifluoromethanesulfonate (NaOTf) ($NaCF_3SO_3$)		−7.5
Sodium bis(fluorosulfonyl)imide (NaFSI)		−8.66
Sodium bis(trifluoromethanesulfonate)imide (NaTFSI)		−8.6

Electrolytes for room-temperature Na-S batteries

and the Tokyo Electric Power Company mass-produced HT Na-S batteries and commercialized them in 2003 [25]. Although the HT Na-S battery reached the commercialization stages, the technology was scrutinized as it risks the safety and cost of maintenance at high operational temperatures. The main limitations are the vigorous reaction of molten sodium metal and risks associated with the possibility of a thermal runaway due to minor crack formation on BASE and the corrosive nature of polysulfide with current collectors [26]. Increment in cell temperature with electrochemical reaction and the release of high enthalpy (~−420 kJ mol^{-1}) also contributed to creating risk factor [27]. These factors limited the widespread application of HT-Na-S battery. Since the high-temperature operation limits broader applications and imposes severe safety challenges, attempts were made to lower the operating temperature. The transition from HT-Na-S batteries to intermediate Na-S (IMT-Na-S) batteries, with applications in aerospace sectors, was reported from 1980 to 1990 [28].

As an alternative to HT Na-S battery technology, the IM Na-S battery operated between 120°C and 300°C while keeping a similar chemistry as for HT Na-S

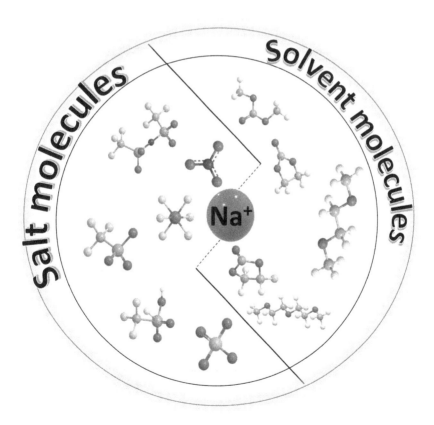

FIGURE 4.1 Overview of various sodium salts and solvent molecules reported for the preparation of liquid electrolyte for RT Na-S batteries.

battery was developed by Abraham et al. and by the US National Aeronautics and Space Administration (NASA) [29,30]. Another significant milestone was the development of the Na-S battery operated at 90°C, which used polymer electrolyte consisting of polyethylene oxide (PEO) in 2006 [31]. Followed by this milestone, Wang et al. developed an RT Na-S battery based on liquid electrolyte (LE) [32]. The electrolyte ystem consisted of sodium perchlorate ($NaClO_4$) salt and a co-solvent of linear/cyclic carbonates (i.e., ethylene carbonate (EC) and dimethyl carbonate (DMC) solvents). In continuation with the developments in LE, ether-based solvents were explored, such as cyclic ethers and linear ethers for RT Na-S batteries [33–36]. Ryu et al. successfully demonstrated sodium trifluoromethane sulfonate ($NaCF_3SO_3$) dissolved tetraethylene glycol dimethyl ether (TEGDME) as an electrolyte for RT Na-S with the room-temperature ionic conductivity of 1.14×10^{-3} S cm^{-1} [37], which is at least one order higher than that of the conductivity of BASE at 300°C. A general overview of various sodium salt and solvent molecule complexes is shown in Figure 4.1.

Figure 4.2 schematically illustrates electrolytes used for RT Na-S batteries. The present chapter aims to provide a holistic overview of the electrolytes for improving RT Na-S batteries. The sodium ion (Na^+) solvation is critical in determining the electrochemical performance of the Na-S battery. The role of electrolytes in enhancing the interfacial properties and their effect on battery performance are discussed comprehensively.

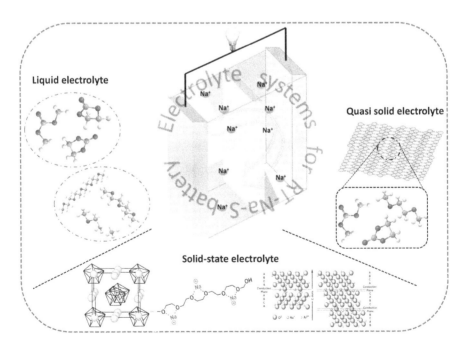

FIGURE 4.2 A schematic illustration of the electrolytes used for RT Na-S battery systems.

4.2 THE SOLID-ELECTROLYTE INTERPHASE (SEI) AND THE IMPORTANCE OF SODIUM-ION SOLVATION

Despite the high theoretical specific capacity of sodium metal with a value of 1166 mAh g^{-1} and its high natural abundance, severe bottlenecks are associated with the development of SMB [38]. The research of SMBs is plagued by uncontrolled dendrite formation and unstable solid-electrolyte interface (SEI). Electrochemical stability and ionic transportation of Na$^+$ are directly affected by the chemical and mechanical properties and even the non-uniformity in the thickness of SEI. Depending on the nature of SEI, stability, reversibility, and Coulombic efficiency (CE) can be modulated. Ideally, a favorable SEI layer permits ionic mobility and restricts electronic conduction. Ideal SEI protects the metal anode from a further parasitic reaction, especially non-reversible polysulfide formation and its shuttling effect in Na-S battery chemistry. However, the formation of an unstable SEI deteriorates the cell performance by catalyzing the non-uniformity in sodium deposition leading to dendritic growth.

Engineering at the electrode-electrolyte interface (EEI) through novel electrolyte additives, solvents, and salt can enhance performance. Ionic transport through SEI depends on the solvation-de-solvation dynamics [39]. Solvation signifies the interaction between solute and solvent in an electrolyte solution [40]. Solvation is accompanied by the formation of secondary bonds, such as van der Waal and hydrogen bonds [19]. Na$^+$ ions form complexes with solvent molecules to stabilize the solvent matrix in the solvated state. Coordination number and stability are the main parameters determining solvation [41].

Furthermore, the dynamics of solvation and de-solvation are determined by the ion-ion and ion-solvent interactions and the accumulated ions at the electrodes [42]. Ion-solvent interaction is a factor that governs the dynamics of solvation and modifies the viscosity and ionic conductivity of an electrolyte. The solvation process/hydration process (defined for aqueous media) occurs when solvent and salt ions interact. As solvent surrounds the salt ions, a diffuse multi-shell network develops. This process is described as solvation [43]. Solvation number is also a related term defined as the density of solvent molecules in a solvation shell. The solvation number is controlled by factors such as the dielectric permittivity of the electrolyte and ionic size [44].

Another factor determining electrolyte properties is chemical polarity. It determined the solubility and solvation strength of the solvent [11]. Predictions using theoretical methods, such as density functional theory (DFT) and molecular dynamics (MD), show that in the classification of carbonate solvents, EC displays high free energy for solvating Na$^+$ [45]. However, EC being solid at room temperature and melting at 36°C requires a co-solvent, usually propylene carbonate (PC) or DMC. Moreover, EC is highly polar with a dielectric constant value of 89.78 at a temperature of 40°C and interacts through dipole interactions with salt-ions, making tetrahedral coordination around Na$^+$ functions as one of the ideal solvents for SMBs [45,46]. While for ether-based solvents, the solvation behavior of Na$^+$ tends to be sluggish as ether possesses low solvent polarity and solvation dynamics [47].

Reduction of electrolytes at EEI can be favorable for the formation of SEI. Solvent reduction leads to the formation of radical species, which results in a thin coating over the electrode. The experimental techniques and theoretical tools can be used in conjunction to analyze electrolyte degradation. Chen et al. [18] found that gas evolution reaction at sodium anode is due to the formation of complexes between Na^+ and carbonate-based electrolyte solvent, thereby lowering the LUMO energy level. From ab initio molecular dynamics (AIMD), it was seen that at C1–O3 and C3–O3 positions, C-O bonds were broken in PC solvent, contributing to carbon monoxide and propane gases forming. Analysis of ion-solvent compatibility was examined by Shakourian-Fard et al. [48] through a quantum chemistry approach. Interactions between Na^+ ions and carbonyl groups of carbonate solvent molecules contributed to shifting LUMO and highest occupied molecular orbital (HOMO) toward lower values. Exchanges were made possible by the carbonyl group, which consisted of an oxygen atom with a lone pair that showed preferential interaction with solvated Na^+ ions. Wahlers et al. [49] investigated the effect of sodium salt concentration and solvation for ether-based solvents. Ion pair formation was evident from the Fourier transform IR (FTIR) spectroscopy analysis and MD simulations at a low concentration of 0.5 M. Aggregation effects between triflate anions and Na^+ were profoundly shown as Na^+ ions concentration exceeded 1.5 M.

4.3 LIQUID ELECTROLYTE TO QUASI-SOLID-STATE ELECTROLYTE TO SOLID ELECTROLYTE FOR SODIUM-SULFUR BATTERIES

Based on the type of electrolyte used, a battery can be called a liquid state, gel-polymer or quasi-solid-state, or all-solid-state battery. Compared to LIB technology, the RT Na-S technology is in its infancy, and developments are nascent. Although HT Na-S batteries employed ceramic solid-state electrolyte (SSE), namely beta-alumina solid electrolyte (BASE), the development of LE catalyzed the research momentum of RT Na-S batteries. Advantages of RT Na-S chemistry include its higher theoretical energy density (1274 Wh kg^{-1}), as it ensures transformation from long-chain polysulfides to sodium sulfide (Na_2S), and improved safety [50,51]. Despite that, polysulfide dissolution into LE and its shuttling cause a gradual capacity decay, and multifaced strategies were explored to mitigate the issue. Engineered cathodes with tailored electrolytes have directly rectified capacity decay. In particular, from a material perspective, electrolyte engineering by exploring various electrolyte systems, such as quasi-solid-state and SSEs, has shown promising performance. Table 4.3 lists different Na-S battery systems and their electrolyte composition and respective electrochemical properties.

Considering the development of RT Na-S, in 2007, a carbonate-based electrolyte system was introduced, which transformed the high-temperature Na-S chemistry to function at room temperature [32]. This was one of the initial works on LE for Na-S battery systems. Followed by the development of carbonate-based LE, such as EC and DMC, DEC and PC were also introduced [32,59,60]. The applicability of ether-based electrolyte solvents, including cyclic and linear molecules, was also reported, which showed high ionic conduction of ~ 10^{-3} S cm^{-1} [33–37]. LEs provide good interfacial properties with electrodes and ensure high ionic conduction research

TABLE 4.3
Various Na-S battery technology, electrolyte composition, and its respective electrochemical properties

Na-S battery classification	Electrolyte composition	Ionic conduction, reported temperature (S cm^{-1}, °C)	Electrochemical stability window (V)	Cycles	Coulombic efficiency (%)
High-temperature Na-S	Beta-alumina, as a solid-state electrolyte [21]	0.1 to 1, 350°C–300°C	2.08 and 1.78	4500	>98
RT Na-S with liquid electrolyte	1 M NaCF$_3$SO$_3$ in diglyme [52]	~3 × 10–3, 25°C	0.6–2.6 vs. Na/Na$^+$	500	~100
	1 M NaClO$_4$ in EC/DMC/PC (1:1:1) [53]	~5–8 × 10–3, 25°C	0.6–2.6	500	~100
Solid state RT Na-S	1 M NaCF$_3$SO$_3$ in diglyme [54]	0.22 × 10–3, 25°C	0.6–2.6	500	99.7
	Na$_3$PS$_4$ [55]	1.43 × 10–4, 25°C, 3.45 × 10–4, 60°C	–	50	~100
	Na$_{3.1}$Zr$_{1.95}$Mg$_{0.05}$Si$_2$PO$_{12}$ [56]	3.5 × 10–3, 25°C	4.5	100	~100 (except the first cycle)
	PEO-NaCF$_3$SO$_3$ (weight ratio as 9:1) [31]	3.38 × 10–4, 90°C	–	10	–
	PEO-NaFSI-TiO$_2$ [57]	4.89 × 10–4, 60°C	4.31	100	~100
	PEO-NaCF$_3$SO$_3$-MIL-53(Al) [58]	6.87 × 10–5, 60°C and 6.52 × 10–4, 100°C	–	50	~100

in LEs proliferated. Functional additives for LEs may include flame-retardant compounds, alloying compounds, etc., which are compatible with LEs [61,62], which were difficult to imagine for solid electrolytes. Electrolyte additives that can alter the physicochemical properties of the SEI, reduce, decompose, or even polymerize over the metal anode, thereby tailoring various properties, have been reported in recent years. Among various additives, fluorinated compounds and their derivatives are among the most explored systems. Inorganic-rich SEI was formed by incorporating HF and H_2O into an ether-based electrolyte, which showed dendrite-free morphology at the anode [11]. However, developing a thicker SEI increases the charge transfer resistance observed in the sodium metal-based symmetric cell. Wu et al. formulated LE-based RT Na-S based on trimethyl phosphate and FEC with NaTFSI salt as an effective electrolyte showing flame retardancy and stable cycling performance [63]. Ionic conductivity measurement showed that 2 M NaTFSI in trimethyl phosphate and FEC electrolyte exhibited a high conduction value of about 6×10^{-3} S cm^{-1} with a wider ESW of ~4.6 V vs. Na/Na$^+$.

Moreover, a full cell fabricated with sulfurized poly acrylonitrile (SPAN) based cathode at 1 C delivered a capacity of 788 mAh g^{-1} after 300 cycles, with stable cycling with a negligible capacity decay. Generally, the FEC inclusion in the electrolyte contributes to forming a stable SEI due to the preferential reduction of the fluorine functional groups on the sodium metal surface. The experimental findings agree with the *ab initio* molecular dynamics simulations.

Although LE-based RT Na-S battery has been investigated in the past, several issues still exist that need immediate rectification. In RT Na-S batteries, LE is accountable for accelerating the loss of active materials and shuttling of the discharge species. Besides tailoring the electrolyte, cathode engineering has also been considered to restrict the loss of active material and limit the polysulfide shuttling effect [64].

As good ionic conduction with a high degree of wettability ensured the mobility of ions, serious issues such as dendritic growth, shuttling effects of polysulfides, and its irreversible dissolution in LE led to capacity decay. Furthermore, considering safety as an essential prime factor, issues such as electrolyte leakage and solvent flammability impeded using LE as an industrially viable system [65,66]. To rectify the challenges of LEs, developments are triggered for a concept referred to as electrolytes beyond LEs. The major candidates for this system are the high concentrated electrolytes (HCEs), localized high-concentrated electrolytes (LHCEs), quasi-solid-state electrolytes/gel polymer electrolytes (GPE), and SSEs, including inorganic and organic SSEs.

4.3.1 Advanced liquid electrolytes for RT Na-S batteries

The conventional LEs for rechargeable batteries, for example, LIBs, have a salt concentration of 1 M, where the recipe functions satisfactorily for ionic mobility and electrode wettability, advanced Les, such as HCEs and LHCEs are unique in their salt composition [67]. For alkali metal batteries, the fresh metal anode exposed to the LE environment is prone to react with the anode repeatedly without a stable SEI. These reactions impede the kinetics, affecting the Coulombic efficiency by irreversible consumption of metal anode and electrolyte inventory. Moreover, in RT

Na/S batteries, the shuttling effect of polysulfides and their dissolution into the LEs promotes capacity decay. The solvent molecules not participating in the solvation process of the cationic or anionic components are the primary source of polysulfide dissolution and shuttling of the discharge products. Therefore, the concept of HCEs is brought into the picture to eliminate the solvent molecules not involved in the solvation process. The HCEs, which generally employ a high salt concentration (above 4 M), can stop the dissolution of polysulfides (due to lack of free solvent molecules) and unwarranted side reactions with the metal anode and, therefore, help elevate the cycle-life of the cell [68]. Salt concentrations above normal levels reduce the availability of free solvent molecules and mitigate the possibility of parasitic side reactions with electrodes. Moreover, from the thermodynamic point of view, anode complexes prefer higher binding with solvent molecules at a high salt concentration and limit their irreversibility solubility [69].

The physical and chemical properties of the HCEs differ significantly from conventional LEs with a salt concentration of 1 M. For example, coordination exists between cations and all the solvent moieties for HCEs, forming aggregates called crystal ion pairs (AGG) [68]. In conventional LEs, solvent-separated ion pairs (SSIP) and contact ion pairs (CIP) are more common. Figure 4.3 shows the schematic representation of various ion complex formations between solvent anions and salt molecules. The formation of AGG in HCEs limits the dissolution of active materials and current collectors and lowers toxicity and volatile nature; however, low ionic conduction and high viscosity are the bottlenecks. Thermodynamic stability and chemical compatibility of salt and solvent at a high salt concentration must be considered while designing and developing HCEs. Viscosity and wettability with separator and electrodes need to be characterized while formulating HCE [67]. Generally, if the solvent is non-compatible with salt molecules at a salt concentration above 4 M, salt crystallization can occur [70]. Literature regarding suitable salt-solvent combinations has been well-reported by Henderson et al. and Brouillette et al. [71,72].

In SMBs, HCEs have been recently studied with salts such as sodium bis(fluorosulfonyl)imide (NaFSI) and sodium bis(fluorosulfonyl)amide (NaN$(SO_2F)_2$) in ether-based solvents, including monoglyme (DME), diglyme, and tetrahydrofuran (THF) [73,74]. Cao et al. employed HCE based on 4 M NaFSI in DME for SMBs, which showed passivation with metal anode and exhibited dendrite-free morphology upon cycling [74]. Figure 4.4 shows a scanning electron microscopy (SEM) micrograph of the cycled sodium metal anode in HCE. Sodium metal over copper after the first plating cycle showed an interwoven mat with a flat geometry similar to pristine sodium before cycling. Upon further cycling, a surface coating was demonstrated on the sodium metal and copper electrode (Figure 4.4c, d). Schafzahl et al. reported a dendrite-free sodium metal anode in NaFSI and DME with a CE of about 98% for 300 cycles [75]. In another work, Lee et al. investigated the applicability of 5 M NaFSI in DME, which outperformed conventional 1 M electrolyte [76]. HCE showed higher viscosity than 1 M NaFSI in DME electrolytes with reduced ionic conductivity. Furthermore, HCE exhibited better electrochemical performance with a CE of about 99.3% at the 120th cycle for sodium plating and stripping in blocking-type cell configuration (i.e., sodium-stainless steel cell). Ion-dipole interactions between NaFSI salt and DME solvent were analyzed through FTIR. A new peak at 1085 cm^{-1} indicating the bonding between

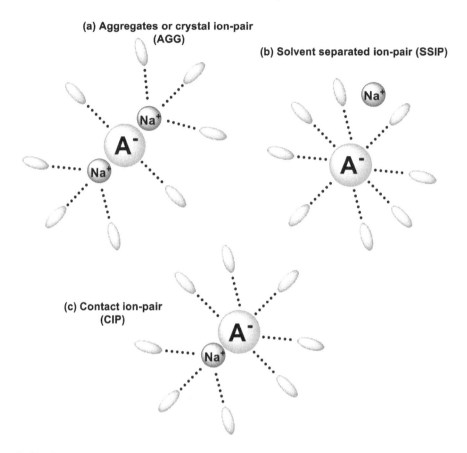

FIGURE 4.3 Schematic representation of various ion complex formations between solvent anions, represented as A- and sodium cations (Na+) (a) aggregates or crystal ion pair (AGG), (b) solvent-separated ion pair (SSIP), and (c) contact ion pair (CIP).

C-O-C linkages with Na+ ions showed an increasing trend with an increase in NaFSI content (Figure 4.4g). The oxidation durability of NaFSI-DME electrolytes could be improved up to 4.9 V (Figure 4.4h), almost double that observed in the conventional LEs. The cell cycled with sodium metal anode, and HCE showed it reduced corrosion of the metal anode with high oxidation durability (above 4.9 V vs. Na/Na+).

Despite all merits of HCEs, there are a few demerits of the high concentration system, For instance, high viscosity, low ionic conduction, and lower wettability with separator along with its high cost (proportional to the salt content in the electrolyte) [76]. Hence, a newer concept of modified HCEs was introduced and termed the Localized high-concentration electrolyte (LHCE) [77]. It was shown that adding the inert solvent (e.g., hydro-fluoro-ether) in the parent solvent can reduce the viscosity without compromising the performance of an HCE [67]. Besides that, LHCEs are unique in their solvation structure with high-concentration salt and solvent clusters distributed in a non-solvating diluent/solvent environment [78]. Thus, for the LHCE, there are mainly three essential components, ionically active salt molecules, solvating

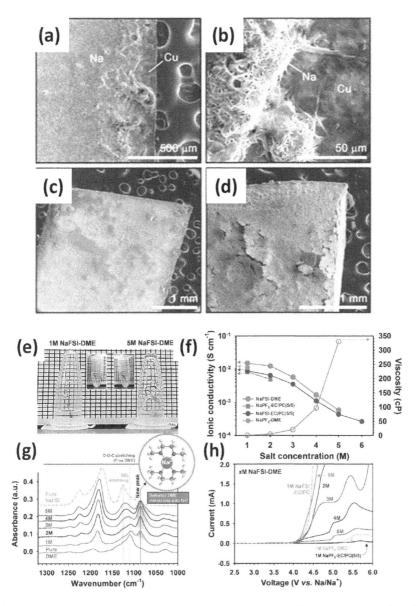

FIGURE 4.4 (a-d) Scanning electron microscopy (SEM), of sodium anode, cycled in HCE (4 M NaFSI-DME electrolyte). (a) Na metal was placed over Cu after the initial plating cycle. (b) A magnified image is shown in (a. c), metallic sodium anode on Cu working electrode after 5th cycle plating. (d) Na metal counter electrode in working electrode after 5th plating cycle. From [74] with permission of Elsevier. (e) Photograph showing 1 and 5 M NaFSI-DME (f) Ionic conductivity and viscosity of various concentrations of NaFSI salt in DME, sodium hexafluorophosphate (NaPF6) in EC/PC, NaFSI in EC/PC, and NaPF6 in DME (g) FTIR spectra of various concentrations of NaFSI in DME (h) Anodic limits of conventional dilute and highly concentrated electrolytes on an Al working electrode at a scan rate of 1 mVs−1. From [76] with permission from the American Chemical Society.

solvent molecules, and a non-solvating inert diluent molecule. The formation of HCE is the foundational step in the preparation of LHCE. Hence, a highly soluble solute and solvent pair are considered, and an HCE is prepared [79]. Including inert diluent does not significantly affect the cation-anion solvation structure, which exists as AGGs. Inert diluent enhances ionic conduction, improves wettability, and lowers the salt concentration in the electrolyte. Generally, hydro-fluoro-ethers function as inert solvents mainly due to their lower dielectric constant and inability to solvate the cations [80]. However, considering the solvent requirements to work on HCE or LHCE is similar to conventional dilute 1 M solvents. This includes high values of dielectric constant, inertness with current collector and other cell components, good wettability, and low viscosity thereby ensuring ionic mobility and wide temperature range [81].

Although the concept of LHCE is promising and explored in LIBs and lithium metal batteries, understanding the chemistry of LHCE in RT Na-S is in its infancy. A few investigations have been reported for RT Na/S batteries. Zheng et al. examined 2.1 M NaFSI in DME with a linear fluorometer, bis(2,2,2-trifluoromethyl) ether (BTFE), as an inert diluent for SMBs [82]. The HCE was initially prepared with 5.2 M NaFSI in DME, followed by dilution with BTFE. Dilution ensured the absence of characteristic bands of DME in LHCE and exhibited the presence of Na^+ and FSI^- AGGs (Figure 4.5), suggesting the solvation structure between Na^+ and DME remains intact. Furthermore, an understanding of LHCE's solvation structure was confirmed through the consistency of a band at ~835 cm^{-1} (Figure 4.5). The inclusion of BTFE enhanced ionic conduction and lowered the viscous nature of HCE, which improved wettability with the separator, thereby enhancing the performance of SMB (Figure 4.5).

Interface engineering of the sodium metal anode through LHCE was examined by Wang et al. using both experimental and theoretical tools [83]. The inclusion of an inert diluent, e.g., 1,1,2,2-tetrafluoroethyl-2,2,3,3-tetrafluoropropylether (TTE) in HCE, which consists of NaFSI in DME, resulted in the formation of LHCE. It is noteworthy to highlight that the diluent decomposes over the surface of the sodium metal anode, which resulted in stabilizing the SEI. Owing to the shift of the HOMO levels from solvent molecules to anions, the SEI developed for the LHCEs was often identified to be rich in inorganic components. The inorganic-rich SEI contributes to the low surface impedance, and good mechanical attributes ensure smoother ionic mobility throughout the interface. In RT Na-S battery applicability of LHCE was investigated by He et al. for the first time [84]. LHCE was formulated with NaFSI in DME solvent with TTE inert diluent. The formulation comprises constituents in a molar ratio of 1:1.2:1 (DME:NaFSI: TTE), which showed a high ionic conduction of about 1.98×10^{-3} S cm^{-1}. Sodium symmetric cell cycled with LHCE showed a low overpotential of 26 mV and exhibited remarkable cycle life for over 590 h. Full cells with sulfur-carbon composite cathode delivered a stable capacity with a low-capacity decay of about 0.10% per cycle. For RT Na-S cells, the comparative performance analysis of the LHCE with conventional LE is depicted in Figure 4.5.

4.3.2 Quasi-Solid-State Electrolytes

Gel polymer electrolyte (GPE) consists of a polymer host matrix and a solvated sodium salt, mobilizing salt with plasticizers and additives. The polymer host matrix mechanically supports the system while the plasticizer participates in ionic mobility. Hence, overall system facilitates ionic conduction and mechanical stability, making it eligible for flexible energy storage devices and innovative wearable technologies. A GPE is fabricated using techniques such as solution casting, phase inversion technique, and electrospinning and then activating by soaking in a LE [85,86]. As polymer hosts can be engineered by co-polymerization, blending, grafting, crosslinking, and filler incorporation, possibilities for tailoring polymer ensure an enormous scope [87–93]. Research in quasi-solid state electrolyte or Gel polymer electrolyte (GPE) in RT Na-S batteries was initiated by Park et al. [94]. Table 4.4 overviews polymer-based electrolytes with their formulation and electrochemical performance for various SMBs.

In the initial work by Park et al., where GPE based RT Na/S cell was fabricated using polyvinylidene fluoride-hexafluoropropylene (PVDF-HFP) with tetraglyme as a plasticizer and $NaCF_3SO_3$ salt in a ratio of 3:6:1 [94]. The first discharge showed two plateaus at 1.73 and 2.27 V, indicating the stepwise reduction of sulfur.

Zhou et al. synthesized pentaerythritol tetraacrylate (PETEA)-tris[2-(acryloyloxy) ethyl] isocyanurate (THEICTA))-based copolymer by radical polymerization for RT Na-S battery [101]. The host (i.e., polymeric backbone) was activated to a quasi-solid-state electrolyte/GPE by soaking in 1 M NaTFSI in PC/FEC solvent. Notably, the state of ionic conductivity strongly depends on the degree of absorption or the time required to absorb the electrolyte. It plays a significant role in determining the performance of the cell. The membranes showed high room-temperature ionic conductivity of about 3.85×10^{-3} S cm^{-1}. The degree of polymerization can be investigated with the help of FTIR, as the absence of C=C bonds signifies the successful formation of the quasi-solid state polymer matrix (Figure 4.6). A reversible plating and striping of sodium in a sodium symmetric cell configuration, i.e., Na//Na, could be performed for over 300 cycles. RT Na-S with the solid-state electrolyte delivered a high reversible capacity of about 736 mAh g^{-1} after 100 cycles.

Despite a comparable ionic conductivity and high initial capacity, a gradual capacity decay is often observed in the quasi-solid state electrolyte, as reported by Kim et al. and Kumar et al., wherein both have employed PVDF-HFP based polymeric matrix [95,98]. Since LE contributes to about 60% of the quasi-solid polymer electrolyte, the polysulfides have a higher chance of getting dissolved in the liquid counterpart. Moreover, the porous geometry with micrometer-sized pores provides a path for the dissolved polysulfides to shuttle, deteriorating the performance.

4.3.3 Solid-state electrolytes

Contrary to LE and quasi-solid polymer electrolyte-based RT Na/S batteries, discharge reactions in SSE differ significantly because the conduction in the SSE occurs via vacancy defects or hopping. Discharge reaction in LE follows a multi-step process, where the transition of longer-chain sodium polysulfides to sodium sulfide occurs at various voltage ranges. A solid-to-liquid transition occurs at 2.6–2.2 V, and a liquid-to-liquid transition happens at a potential voltage range of 2.2–1.65 V. In sharp contrast, SSEs, single step process proceeds through solid-to-solid state transition at a range of 1.65–0.6 V, according to the equation,

$S_8 + 16Na^+ + 16e^- \rightarrow 8Na_2S$

The ion transport in solids was first introduced by Faraday in 1838 while he was researching the ion transport properties of Silver(II) sulfide (Ag_2S) and Lead(II)

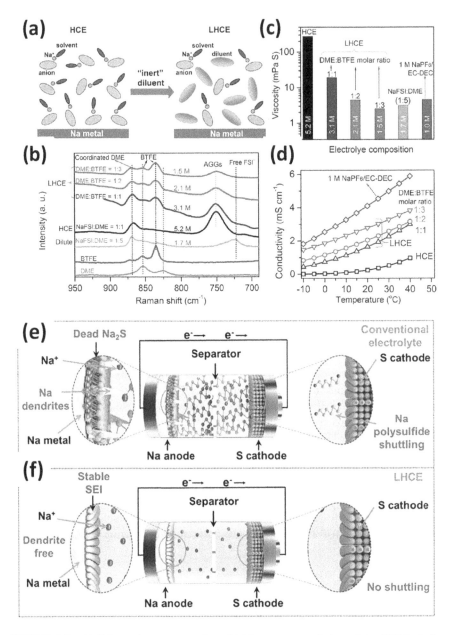

FIGURE 4.5 (a) Schematic representation of high concentrated electrolyte (HCE) design and localized high concentrated electrolyte (LHCE) design in sodium metal batteries. (b) Comparison of Raman spectra for LHCEs and HCEs with varying concentrations of NaFSI (c, d) Viscosity and ionic conductivity of various electrolytes at different temperatures. From [82]/ with permission of American Chemical Society. (e, f) Schematic representation of RT Na-S battery with conventional and LHCE system. From [84]/ with permission of American Chemical Society.

TABLE 4.4
Overview of polymer-based electrolytes in various sodium-based batteries and their electrochemical properties

Sodium battery system	Electrolyte formulation	Ionic conduction, reported temperature (S cm^{-1}, °C)	Electrochemical stability window	Cycles	Coulombic efficiency (%)
RT Na-S	PEO- NaCF$_3$SO$_3$ [31]	3.38×10^{-4}, 90°C	–	10	–
	PEO-NaFSI-TiO$_2$ [57]	4.89×10^{-4}, 60°C	4.31 V	100	~100
	PEO- NaCF$_3$SO$_3$- MIL-53(Al) [58]	6.87×10^{-5}, 60°C and 6.52×10^{-4}, 100°C	–	50	~100
	PVDF- NaCF$_3$SO$_3$- tetraglyme [94]	5.1×10^{-4}, 25°C	–	20	–
	PVDF-HFP-NaTf-SiO$_2$- EC:PC [95]	4.1×10^{-3}, 25°C	~4 V	8	–
	PVDF-HFP NaCF$_3$SO$_3$ EMITf and EC:PC [96]	8.4×10^{-3}	~2.3 V to 2.3 V (4.6 V)	10	–
	PVDF-HFP NaCF$_3$SO$_3$ EMITf [97]	5.7×10^{-3}	~2.4 V to + 2.4 V (4.8 V)	10	–
	PVDF-HFP-NaCF$_3$SO$_3$-Tetraglyme [98]	–	1–3 V	20	–
	PAN-NaCF$_3$SO$_3$–PEGDME [99]	~10^{-3}, 30°C	1–2.5 V	40	–
	Poly-PETA-NaTPB-PC/FEC [100]	2.3×10^{-3}, 25°C	0.5–3 V	2500, 2 C	~100
	PETEA copolymer-NaTFSI-PC/FEC [101]	3.85×10^{-3}, 25°C	0.6–2.6 V	100, 0.1 C	~100
Na-metal battery	PVDF-HFP-GO-NaClO$_4$-EC/PC [102]	2.3×10^{-3}	4.7 V	1100, 1 C	99.8
	PVDF-HFP-β/β″-Al$_2$O$_3$-NaClO$_4$-EC/DEC/FEC [103]	7.13×10^{-4}, 25°C	2.4–3.9V	1000, 1 C	99
	PVDF-HFP-NaPTAB-PC [104]	0.94×10^{-4}, 25°C	5.2 V	500, 0.5 C	~100
	PEO-Cu MOF-NaClO$_4$-EC/DEC/FEC [105]	3.48×10^{-3}	2.4–3.6 V	800, 1 C	~100
	PEO-SiO$_2$-NaClO$_4$-Emim FSI [106]	7×10^{-4}, 25°C	4.2 V	100, 0.05 C	~100
	PEO-PPC-NASICON [107]	0.12×10^{-3}, 25°C	2.4–3.6 V	500, 1 C	~100
	Thermoplastic PU-NaClO$_4$-EC/DEC/FEC [108]	1.5×10^{-3}, 25°C	2.5–4 V	200, 0.25 C	~100
	Crosslinked POSS-PEG-NaClO$_4$ [109]	4.5×10^{-6}, 30°C	4.5 V	50, (15, 30, and 60 mA g^{-1})	~100
	Poly EPTA-NaPF$_6$- PC:EMC:FEC [110]	5.33×10^{-3}, 25°C	5.5 V	1000, 100 mA g^{-1}	98.3
Na-Air/ Na-O$_2$ battery	PEGDME500-NaTFSI-Tetraglyme [111]	–	1.9–4.3 V	–	–

fluoride (PbF$_2$) [112]. Ion diffusion in solids is described by a mechanism known as the hopping mechanism [113,114]. Ions hop from neighboring lattices through the interconnected network in solids [13]. Other mechanisms have also been developed for ionic transport in solids, such as the knock-off and direct-hopping mechanisms. In the knock-off mechanism, an interstitial ion knocks off an adjacent ion in the lattice, and cations are diffused in the interstitial sites of a lattice [44]. The SSE can be classified into inorganic solid-state electrolytes (ISEs) and organic or solid polymer electrolytes (SPE) [115].

4.3.3.1 Inorganic solid-state electrolytes

The ISEs comprise mobile ions that are distributed in a symmetric structure. Point defects in the lattice enable the shifting of mobile ions, thereby facilitating ionic conduction, with values in the range of 0.5–5 × 10^{-4} S cm^{-1} at 25°C [116]. Hence, the factors relating the ionic conduction in ISE include the presence of mobile ions and their energy and vacancies or defects in the matrix. The sulfide and oxide-based electrolytes are the classic example of ISEs. Sulfide-based ISEs have the advantage of low-temperature synthesis with intrinsic ionic conduction of ~0.2 mS cm^{-1}, hence finding its applicability in high-power batteries [117]. The sensitivity toward moisture and toxicity of sulfide base electrolytes limits their widespread applications. The high ionic conduction in sulfide-based ISE can be correlated with the host atoms' larger ionic size and electronegativity. The larger size and low electronegativity of sulfur atoms (atomic radii ~105 pm and electronegativity (Pauling) ~2.58) channelize the ion transport for Na$^+$. Moreover, high elasticity and low resistance at the grain boundary also contribute positively to ionic conduction [118].

Synthesis of sulfide-based ISE involves mechano-chemical techniques [119,120]. Ball milling is found to be an effective method for the reduction of crystallite and particle size. It also enables materials to have micro- and macrostrains in the matrix and contributes to enhancing ionic conductivity at ambient conditions [121]. Trisodium; sulfanylidene(trisulfido)-lambda5-phosphane (Na$_3$PS$_4$) is one of the most reported sulfide-based ISE for RT Na-S batteries. It possesses two structural polymorphs depending on the synthesis conditions tetragonal Na$_3$PS$_4$ (t-Na$_3$PS$_4$) and cubic Na$_3$PS$_4$ (c-Na$_3$PS$_4$) [122,123]. The synthesis method of sodium tetra thiophosphate through a solid-state reaction showed ionic conductivity of c-Na$_3$PS$_4$ as 7.71 × 10^{-4} S cm^{-1} at 250°C and for t-Na$_3$PS$_4$ as 1.55 × 10^{-2} S cm^{-1} at 450°C [124]. Notably, t-Na$_3$PS$_4$ electrolyte is more often reported in the literature [125]. Hayashi et al. synthesized a c-Na$_3$PS$_4$ glass-ceramic electrolyte, which exhibited ionic conductivity as 2 × 10^{-4} S cm^{-1} [126]. The preparation involved the reduction of glass-ceramic grain boundaries, which channeled the ions, facilitating smooth conduction pathways. Ionic conduction can also be tailored by partial substitution of P by Si, which modifies the structural arrangements in c-Na$_3$PS$_4$ [127]. Simulation data recorded through *Ab initio*, molecular dynamics simulations indicated the sodium disorder in the c-Na$_3$PS$_4$ structure as Si is introduced, which modified the ionic conductivity [128,129].

The electrochemical properties of the cell can be engineered by the mechanical process of the ISEs [117,130]. Techniques such as sintering, annealing, mechanical milling, and quenching showed a positive shift of ionic mobility in t-Na$_3$PS$_4$ by expanding the 3D lattice to promote diffusion pathways [131].

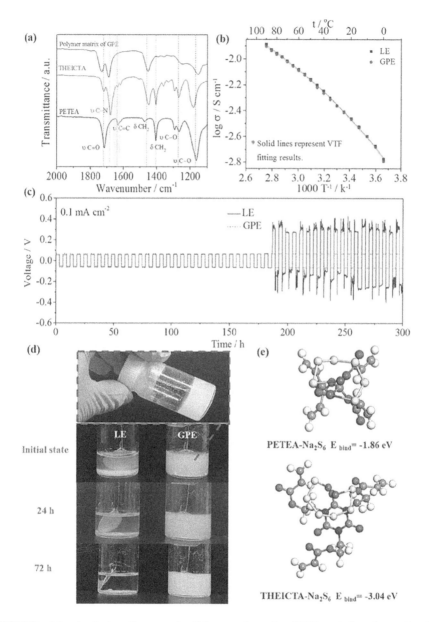

FIGURE 4.6 Analysis of a quasi-solid-state electrolyte/GPE and its electrochemical performance, (a) FTIR spectra of monomer and polymer matrix indicating conversion reaction, (b) temperature dependency of ionic conduction in quasi-solid-state electrolyte/GPE and for LE based (1 M NaTFSI in PC/FEC solvent), (c) Galvanostatic cycling measurement on a Na-Na symmetric cell with GPE and LE (1 M NaTFSI in PC/FEC solvent) at 0.1 mA cm−2 current density, (d) Photographs showing the formation of sodium polysulfides and its dissolution in LE and its comparison with GPE at various time, and (e) binding energy calculation and the molecule bonded with Na2S6, from first principle calculations. From [101]/ with permission of John Wiley & Sons.

The ionic conduction has also been enhanced by defect substitution and doping [128,132]. For instance, creating 2% vacancies in the Na_3PS_4 crystal lattice improved the ionic conduction by 0.2 S cm^{-1} [133]. In another work, bromine-doped t-Na_3PS_4 exhibited similar ionic conductivity (2.37 mS cm^{-1}) compared to LE [134]. The purity of precursors also contributes to the electrochemical performance of the solid electrolytes. Moreover, synthesis methods also regulate ionic conduction and related properties [135]. All-solid-state sodium battery with c-Na_3PS_4 ISE prepared from sodium sulfide (Na_2S) and phosphorus pentasulfide (P_2S_5) exhibited room-temperature ionic conductivity as high as 2.6×10^{-6} S cm^{-1} [136]. While Na_2S: P_2S_5 in a weight ratio of 3:1 showed enhanced room-temperature ionic conductivity of 4.2×10^{-4} S cm^{-1} [135].

An initial report on employing sulfide-based ISE in RT Na-S was demonstrated by Nagata et al. in 2014 [137]. By optimizing the ratio of Na_2S: P_2S_5 as 3:1, Na_3PS_4 electrolyte showed room-temperature ionic conductivity of about 0.13 mS cm^{-1}. Moreover, the cell at a constant current density of 0.13 mA cm^{-2} at 25°C delivered a high reversible 1522 mAh g^{-1} capacity. Followed by this work, there were many reports on employing Na_3PS_4 as a Na$^+$ conducting system for SMBs. Na_3PS_4 electrolytes modified by mechanical milling and annealing when used in RT Na-S battery showed high ionic conductivity of 1.09×10^{-4} S cm^{-1} at 28°C with stable cycling [138]. Cathode consisted of carbon nanocomposite with Na_3PS_4-Na_2S, where Na_3PS_4 served a two-fold purpose, i.e., functioned as active material at the catholyte part and ISE. Wan et al. modified sulfide-based ISE using phosphorous-free sodium thioantimoniate (Na_3SbS_4) and employed it as an electrolyte for RT Na-S battery [139]. Nano-scaled electronic/ionic network tailored to the cathode facilitated electron and ion transport. Moreover, through the mechano-chemical process, the cathode and electrolyte enhanced interfacial contact and reduced stress/strain at the interface.

An et al. investigated the role of engineered interfaces for Na_3PS_4-based solid-state RT Na/S battery, where the interface of the sodium metal anode was alloyed by tin. The interface was further activated with an ionic liquid based on N-butyl-N-methylpyrrolidinium bis(fluorosulfonyl)imide (Pyr14FSI) to ensure better adhesion with the ISE [140]. The symmetric cell showed an overpotential of 0.55 V vs. Na/Na$^+$ after 900 h. A full cell of selenium-doped sulfurized polyacrylonitrile-based sulfur cathode exhibited an initial capacity of 708.5 mAh g^{-1}.

Apart from sulfide-based ISE, sodium (NAtrium) SuperIonic CONductor (NASICON) electrolytes and β-alumina solid electrolytes constitute oxide-based ISEs. Chemical stability, moisture resistance, and low toxicity are the most important features of oxide-based ISEs, making them superior to sulfide-based ISEs. The development of NASICON electrolytes was initiated in 1976 by the Novel laureate Prof. John Goodenough [24]. The general formula of NASICON may take the form of $Na_{1+x}Zr_2Si_xP_{3-x}O_{12}$ (where $0 \leq x \leq 3$) [141]. Ambient temperature ionic conductivity varies from 10^{-5}–10^{-2} S cm^{-1}. NASICON could be prepared through co-precipitation, solid-state reaction, sol-gel, and polymerization [142,143]. It is noteworthy to highlight that the ionic conductivity is sensitive toward the synthesis route. Manthiram et al. successfully demonstrated $Na_3Zr_2Si_2PO_{12}$-based NASICON electrolytes to prevent the shuttling effect of polysulfides and addressed the limitation of the resistive anode/electrolyte interface [144]. NASICON membrane coated with an intrinsic nano-porous polymer and wetted with a few drops of 1 M NaClO$_4$ in the

TEGDME electrolyte exhibited an ionic conductivity of 4.1×10^{-5} S cm^{-1}. Aluminum doped monolithic ISE with $Na_{3.4}Zr_{1.9}Al_{0.1}Si_{2.4}P_{0.6}O_{12}$ structure showed high ionic conductivity of 4.43×10^{-3} S cm^{-1} at 50°C [145]. Aluminum doping improves the unit cell volume, enabling ionic conductivity augmentation. IM Na-S cell employing $Na_{3.4}Zr_{1.9}Al_{0.1}Si_{2.4}P_{0.6}O_{12}$ electrolyte showed the Coulombic efficiency of about 99.9%., with a specific discharge capacity of about 300 mAh g^{-1} even after 480 cycles.

Sodium-β/β″-alumina is another exciting candidate for oxide-based ISEs. Initial works on sodium-β/β″-alumina were focused on HT Na-S batteries [146]. Temperature catalyzes ionic conductivity in β″-alumina electrolytes, thus making it suitable to conduct Na$^+$-ion, and it was employed for the HT Na-S system [147]. Microstructure modifications by doping, for instance, manganese dioxide, silicon dioxide, magnesium oxide, and zirconium dioxide, and tuning of the ratio of β″-alumina and β-alumina the ionic conductivity of the electrolyte could be enhanced by one order of magnitude [148–150].

4.3.3.2 Solid polymer electrolyte

SPE is one of the widely studied materials based on its versatile nature and considering its potential application [151]. SPE constitutes the organic-based SSE, which consists of Na salt incorporated macromolecules as Na$^+$ conduction membrane [16]. The appealing features of SPE include good mechanical properties, low density, high safety, reduced side reactions, thermal and chemical stability, and roll-to-roll processability. Furthermore, the added advantage of SPE includes low volatility, leak-proof and solvent-free system, with high mechanical, thermal, chemical, volumetric, and electrochemical stability [152]. Although SSEs constitute a broad spectrum consisting of ISEs, which can be again bifurcated into oxide-based and sulfide-based electrolytes, SPEs provide the unique combined features of tailoring ionic conduction, enhanced mechanical flexibility, and cost-effectiveness than the ceramic and glassy electrolyte.

To date, various sodium salts have been studied to dissociate in several polymer matrices, thereby facilitating the availability of Na$^+$ toward high ionic conductivity. The presence of polar functional groups catalyzes salt dissolution and allows cationic transportation. Ion transport in SPE is dependent on the ability of Na$^+$ to move between the ion coordination sites [153,154]. Besides the availability of coordination sites, temperature also influences ionic conductivity in polymer electrolytes [155]. Various theoretical models can correlate the dependency of temperature on ionic conduction. These can be the Arrhenius, Vogel-Tammann-Fulcher (VTF) models [156]. Ion hopping mechanisms along coordination sites and solvating sites, for instance, the presence of O atoms or N atoms in the host matrix, are well reported in recent review works [116,157,158]. As macromolecules have semi-crystalline nature with a mix of domains consisting of crystalline and amorphous parts, ionic conductivity has been more pronounced about the amorphous part in the polymer. As amorphous parts are less arranged and flexible, ionic mobility is well facilitated by amorphous regions.

Strategies to reduce crystalline phases have positively enhanced ionic mobility, as its amorphous nature is correlated with chain mobility [159]. Thus, approaches have been identified for reducing the crystalline nature of SPEs [160]. This can be achieved by plasticizer incorporation, copolymerization, blending, and adding fillers [161–163]. Various blend systems, copolymers, and nano/micro filler-reinforced matrices have been developed to enhance ionic conductivity. Furthermore, ionic conductivity is also dependent on salt concentration in the macromolecule, which influences the segmental motion of the polymer chain [164]. Glass transition temperature (T_g), which is related segmental motion of a polymer, thus has a vital role in ionic transport [153,165,166].

Amongst polymer hosts for SPEs, polyethylene oxide (PEO) is one of the most explored SPEs for battery applications, where its linear structure facilitates ionic transportation [157,167,168]. PEO consists of oligoether units, where ether-oxygen linkages contribute positively to ion mobility by its segmental motion [169–172]. Apart from PEO, linear polymers such as copolymers of PVDF, polyvinylpyrrolidone (PVP), poly methyl methacrylate (PMMA), polyvinyl alcohol (PVA), etc., also showed Na$^+$ conduction [97,146,154]. In addition to the advantage of processing, with mechanical and thermal properties, SPEs face several limitations which need to be addressed, considering their industrial viability. In particular, the issues of low ionic conductivity of SPE limit its practical usage in real-world applications in which the device works at room temperature and below. Furthermore, electrode-electrolyte intimacy and high interfacial resistance limits cycling and performance of battery. Hence considering the demand for safer and leak free electrolyte system, rectifications are utmost important in order to escalate the realization toward commercially acceptable electrolyte system.

Considering initial works on Na-S battery with SPEs, PEO with NaCF$_3$SO$_3$ was reported to have good reversibility. The cell was operated at 90°C with its ionic conductivity as 3.38×10^{-4} S cm^{-1} [31]. The cell exhibited the first discharge capacity of 505 mAh g^{-1} sulfur. However, decay in capacity was evident with its values after ten cycles as 166 mAh g^{-1} sulfur. As mentioned, nanofillers are incorporated in SPE to enhance ionic conductivity by improving the amorphous nature in the host matrix. Aluminum-based MOF (MIL-53(Al)) as a nanofiller showed promising results, considering its performance PEO- NaCF$_3$SO$_3$ based all-solid-state RT Na-S battery [58]. Filler incorporation enhanced the segmental motion of PEO with the highest ionic conductivity reported as 6.87×10^{-5} S cm^{-1} at 60°C and 6.52×10^{-4} S cm^{-1} at 100°C. It is to be highlighted that the best ionic conduction was observed when EO:Na mole ratio was maintained at 20 with 3.24 wt.% of MIL-53(Al) nanofiller. In another work, the addition of nanofiller based on titanium dioxide (TiO$_2$) at 1% to PEO matrix with sodium bis(fluorosulfonyl)-imide (NaFSI) showed enhancement in ESW as 4.31 V (vs. Na$^+$/Na)) with a high Na$^+$ transport number (t_{Na}^+) of 0.394 [57]. The prepared SPE exhibited ionic conductivity of 4.89×10^{-4} S cm^{-1} at 60°C and found applicability in all-solid-state Na-S battery. In addition to a reduction in crystallinity, differential scanning calorimetry (DSC) showed a decrement in the crystalline melting temperature of the SPE (Figure 4.7). Furthermore, full cell analysis revealed that the fabricated all-solid-state Na-S cell, operated at 60°C, delivered CE as ~100% with a stable discharge capacity.

As filler incorporation has a positive effect on enhancing ionic conductivity and widening ESW, another approach to improve the performance of SPE is by blending technique [173]. Reduction of crystallinity with enhancement in ionic mobility was achieved by combining approach, which is well reported for LIB [174–177]. Saroja et al. blended PVDF-HFP and (poly (butyl methacrylate)) (PBMA) and incorporated white graphene (boron nitride), which functioned as Na$^+$ selective membrane for RT

Na-S battery [178]. Room-temperature ionic conductivity for the optimized blend nanocomposite membrane was shown as 1.134×10^{-3} S cm^{-1}. It is noteworthy to highlight that the membrane also acted as a polysulfide shielding layer. Various analyses identified that boron and nitrogen atoms in boron nitride resisted the shuttling process. In Na-S full cell, a capacity retention of about 83.1% after 500 charge-discharge cycles was shown when optimized blend nanocomposite was employed.

In summary, SPEs evolved as a suitable candidate to mitigate polysulfide dissolution and suppress the shuttling effect of sodium polysulfides. Although the research in SPEs is in its infancy, efforts are being devoted to developing high ionic conductivity, low interfacial resistance, and a broader operating stability window for SPEs. Exploration of various functional fillers such as boron nitride reported by Saroja et al. [178], which can preferentially absorb and trap polysulfides, thereby mitigating

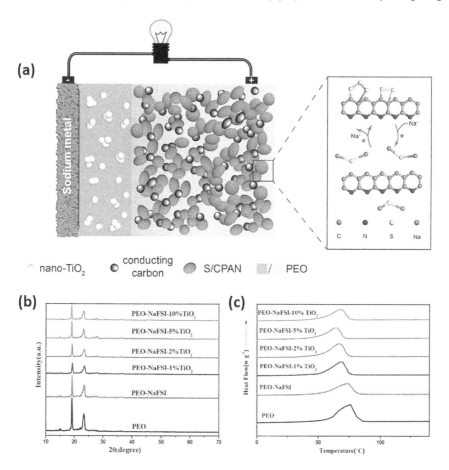

FIGURE 4.7 (a) Schematic illustration of all-solid-state Na-S battery consisting of the cathode (sulfurized-poly acrylonitrile (S-PAN)), SPE based on PEO-NaFSI-TiO2 electrolyte and Na metal anode, (b) Overlaid XRD patterns of PEO, PEO-NaFSI, and PEO-NaFSI with TiO2 at varying amounts (c) DSC thermograms of various PEO-NaFSI-TiO2 composite. From [57]/ with permission of ACS Publications.

polysulfide shuttling, can be investigated for further development of SPEs. Hence by judicious selection of functional fillers, plasticizers, and additives, research in SPEs can be further escalated considering their industrial advantages.

4.4 NATURE OF ELECTROLYTE AND ITS ROLE IN DEVELOPING A STABLE SEI AND CEI

Since electrode/electrolyte interphase is critical in ensuring long-term cycling stability, unveiling the complexities at the interface is of utmost importance. The mutual interaction of the ion and solvent has a pivotal role in determining the nature of the interphase. The interphase is a multi-component complex system consisting of organic and inorganic compounds. The interface differs from the interphase, as the former is a two-dimensional boundary. In contrast, the interphase is a three-dimensional arrangement with a few nanometer-thick layers formed at the surface of an electrode [179]. Since the interphase is a solid arrangement of the organic and inorganic species and can conduct the ions, it is often called "solid electrolyte interphase (SEI)." As discussed previously, the interaction among electrolyte salt, solvent, and additives can tweak the chemical and physical properties of the SEI.

To date, many studies have unraveled the mechanism behind the formation of SEI through the electrochemical or chemical decomposition of electrolyte salt and solvent. Due to the relatively negative reduction potential of sodium, most organic solvents get reduced chemically. Upon cycling, the interphase grows and dissolves in the electrolyte repeatedly. It leads to consuming the sodium metal anode and electrolyte inventory, causing capacity fade and decay in the Coulombic efficiency. However, the relative energy levels for the composite sulfur cathode lie close to the energy levels of the electrolyte; it does not get reduced or oxidized chemically. However, the electrolyte species can repeatedly get reduced, or oxide upon charge/discharge may lead to forming a cathode-electrolyte-interphase (CEI). The formation of CEI is highly debatable due to its high degrees of variability.

The chemical potential of electrolytes governs the formation of SEI and CEI. The redox reaction between electrolyte molecules is initiated as the nucleophilic cathode materials, Lewis's base in nature, interact with electrophilic atoms of electrolyte. In particular, in the case of CEI, the electrolyte oxidizes as its HUMO level is higher than the Fermi energy of the cathode. And for SEI, the electrolyte reduces if the Fermi energy of the anode is above the LUMO of the electrolyte. Hence electrochemical stability window (ESW) of electrolyte functions as a benchmark for SEI and CEI formation. Ideally, for an SMB, the requirements for an ideal EEI are as follows: electronically insulating but ionically conductive, physicochemical stable, and electrochemically active to prevent electrolyte reduction and ensure electrochemical performance. Furthermore, EEI should be mechanically robust and structurally stable, suppressing dendrite growth and withstanding volumetric changes during cycling.

In addition to forming SEI, dendrite growth at the anode also deteriorates cell performance. The formation of SEI directly affects the electrode overpotential and alters the localized current density, which catalyzes the growth of dendrites [180]. Ma et al. investigated electrolyte additives' role in preventing SEI dissolution [181]. It was observed that the electrolyte chemistry, which depends on salt, solvent, and additives, showed a definite role in the composition of SEI, capacity decay with cycles, and overall performance. Electrolyte formulation using additives can alter the interphase, forming a robust interface. For instance, in the case of a carbonate-based electrolyte system, an electrolyte additive that can repair interphase should have a low oxidation potential, thereby ensuring oxidation of additives on the electrode proceeds first. Sun et al. incorporated electrolyte additives based on ionic liquids such as ethylaluminum dichloride and 1-ethyl-3-methylimidazolium bis(fluorosulfonyl)imide, which showed high ionic conduction with non-flammability for SMBs [182]. Two electrolyte additives were investigated: ethylaluminum dichloride and 1-ethyl-3-methylimidazolium bis(fluorosulfonyl)imide. The chemical composition of SEI analyzed using XPS and Cryo-TEM showed the presence of inorganic rich composition with major components such as NaF, NaCl, and Al_2O_3. The introduction of additives showed 90% of capacity retention after 700 cycles, enhancing the performance of SMB.

Mitigating dendrite formation in metal batteries has been focused on improving performance and efficiency. Hence investigators reported nanodiamond particles as an effective electrolyte additive that can reduce lithium dendrite formation [183]. Similar studies for sodium metal anode showed suppression of dendrite formation upon adding functionalized diamantane in diglyme-based liquid electrolyte [184]. Bis-N,N'-propyl-4,9-dicarboxamidediamantane (DCAD) was incorporated in diglyme electrolyte in 0.5 M sodium trifluoromethanesulfonate (NaOTf). Smooth surface morphology observed from SEM analysis after cycling with DCAD additive for 15 cycles and from visualization of sodium deposition with a symmetric Na//Na glass cell of sodium anode proved suppression of dendrite formation. Dendrites were prominent for cells without additives, and the electrode was utterly covered at 420 µAh.

Moreover, incorporating DCAD resulted in stable cyclic charge-discharge properties with enhancement in energy density. Cyclic voltammetry showed electrochemical stability of additive with a range of 1.3 to 3.4 V vs. Na^+/Na. Zheng et al. observed that incorporating stannous chloride ($SnCl_2$) as an electrolyte additive to carbonate electrolyte formed an alloy layer of sodium-tin. Subsequently, a stable SEI layer, which is rich in NaCl, was formed [185]. Hence, the SMB showed high-capacity retention and a reduction in dendritic growth. Deposition of the discharge product on the electrode can negatively affect the performance of the cell by increasing the polarization and fading capacity [186]. Ren et al. employed poly-benzimidazole as a separator and electrolyte additive, sodium iodide (NaI)-phosphorus pentasulfide (P_2S_5). The cell showed dendrite suppression with mitigation of corrosive polysulfide crossover in the electrolyte, thereby contributing to the retention in capacity on the RT Na-S battery [187].

Modifying the anode-electrolyte interface has shown a direct effect on improving the performance of cells [188]. Protecting Na anode by creating a metal-alloy interface (MAI) approach was evaluated for diglyme-based RT Na-S battery by Kumar et al. [54]. Instead of incorporating tin-based additives, which follow a solid-liquid reaction, a solid-vapor reaction was employed by reacting tin tetrachloride vapors to the Na anode. Symmetric cells operated at high current densities of (2–7 mA cm^{-2}) exhibited uniformity in Na deposition with lower overpotential than pristine Na. MAI approach promoted enhanced interface properties, ensured strong adhesion with the anode, and showed suppression of dendrites. RT Na-S cell with MAI-based anode showed stable cycling for more than 500 cycles with CE as 99.4%. In another work, Kumar et al. prepared a biphasic interface juxtaposing NaOH and $NaNH_2$, enabling mechanical properties to the interface and exhibiting stable plating and stripping at high current densities [52]. The full cell cycled with anode tailored using the biphasic interface in 1 M $NaCF_3SO_3$ in diglyme and sulfur-infused microporous carbon cathode showed stable cycling for 500 cycles at 0.5 C.

Furthermore, capacity retention of 55% (of initial discharge capacity) was revealed even after 500 cycles. Soni et al. demonstrated the microarchitecture of CNT grown over stainless steel mesh as a stable host for reversible plating/stripping of Na in 1 M $NaCF_3SO_3$ in a diglyme-based RT Na-S battery [189]. Uniform deposition of Na ensured minimum overpotential, thereby favoring longer and more stable cycling.

Determining the interface requires in situ and operando techniques, which unveil the morphology, compositional details, and physicochemical properties [190]. Such understanding catalyzes the understanding of interfacial properties with the performance property of a cell. Lutz et al. employed ^{23}Na-NMR spectroscopy to investigate SEI over Na metal and derived the Na^+ solvation dynamics [11]. The results indicated that Lewis's basicity of the salt anion is a main factor that regulates the Na^+ solvation for weakly solvating solvents such as ethylene glycol dimethyl ether (DME). Chemical reduction of solvent produces free radicals, which are found to initiate the solvent molecules for polymerization and catalyze the salt anion decomposition.

Furthermore, interpretations from the Nyquist plot for various Na salts in DME demonstrated the stability of SEI. Smaller semicircle formation by the electrolyte containing sodium hexafluorophosphate ($NaPF_6$) salt implied low values of charge-transfer resistance, indicating stable SEI formation. However, electrolytes consisting of $NaClO_4$ and NaOTf exhibited higher values in charge-transfer resistance, indicating unstable SEI formation. In addition to altering the nature of SEI, electrolyte additives can inhibit the dissolution of SEI [181]. Bai et al. evaluated the compositional analysis of SEI formed by cycling in 1 M NaPF6 in DME with vinylene carbonate (VC) and its effect on mechanical properties at the interface [191]. The passivation layer was formed as a result of electrolyte reduction. From XPS analysis, VC reduced to its polymer form as implied from O1s spectra at 534.2 eV [192]. Polymer formation and inorganic-rich SEI contributed to enhancement in the mechanical toughness as probed using AFM. Furthermore, the introduction of VC to DME lowered LUMO's energy level and promoted stable SEI formation.

4.5 FUTURE PROSPECTS

Though RT Na-S battery chemistry is one of the most studied battery chemistries in the last decade, a detailed understanding still requires overcoming inherent challenges. Developing liquid electrolytes for the RT Na-S is an essential milestone in realizing this chemistry to function at ambient conditions. In electrolyte formulation, the selection criteria of sodium salts, solvents, and additives are majorly based on their compatibility with the sodium anode and the sulfur cathode [193]. As metallic sodium can readily react with organic molecules, it can lead to electrolyte decomposition, which adversely affects the cycle life of the cell [194].

The formation of cation/anion solvation shells has a direct relation to regulating the electrochemical properties of the cell [11,195,196]. Electrolyte solvents with high DN (ability to solvate the cationic component of the salt) number endorse the redox reaction of metal polysulfides by shifting kinetic pathways and stabilizing polysulfide anions through multiple states.

With the advancement in data-driven techniques, artificial intelligence, machine learning, and theoretical and computational techniques juxtaposed in various fields, battery technology has benefitted from this development [197]. Incorporating artificial intelligence and machine learning (ML) approaches in developing suitable electrolytes have been initiated to unveil the complexities of ion mobility and its effect on cell performance. For instance, Wheatle et al. employed Bayesian optimization, an ML technique to relate ionic transport and mechanical properties, providing a framework to combine molecular simulation and ML [198]. Thus, theory-guided experimental work can channelize the efforts for developing a more suitable electrolyte system for RT Na-S battery chemistry. Overall, critical evaluation of suitable electrolyte formulations can undoubtedly rectify the issues and realize the development of a practical RT Na-S battery.

REFERENCES

1. Maddukuri S, Malka D, Chae MS, et al. On the Challenge of Large Energy Storage by Electrochemical Devices. *Electrochim Acta.* 2020;354:136771.
2. Dunn B, Kamath H, Tarascon J-M. Electrical Energy Storage for the Grid: A Battery of Choices. *Science* (80-). 2011;334:928–935.
3. Saroha R, Khan TS, Chandra M, et al. Electrochemical Properties of $Na_{0.66}V_4O_{10}$ Nanostructures as Cathode Material in Rechargeable Batteries for Energy Storage Applications. *ACS Omega.* 2019;4:9878–9888.
4. Sungjemmenla, Soni CB, Vineeth SK, et al. Unveiling the Physiochemical Aspects of the Matrix in Improving Sulfur-Loading for Room-Temperature Sodium–Sulfur Batteries. *Mater Adv.* 2021;2:4165–4189.
5. Sundaram PM, Soni CB, Sungjemmenla, et al. Reviving Bipolar Construction to Design and Develop High-Energy Sodium-Ion Batteries. *J Energy Storage.* 2023;63:107139.
6. Soni CB, Sungjemmenla, Vineeth SK, et al. Challenges in Regulating Interfacial-Chemistry of the Sodium-Metal Anode for Room-Temperature Sodium-Sulfur Batteries. *Energy Storage.* 2022;4:e264.
7. Chu H, Noh H, Kim YJ, et al. Achieving Three-Dimensional Lithium Sulfide Growth in Lithium-Sulfur Batteries Using High-Donor-Number Anions. *Nat Commun.* 2019;10:1–12.
8. Gutmann V. Empirical Parameters for Donor and Acceptor Properties of Solvents. *Electrochim Acta.* 1976;21:661–670.
9. Zou Q, Lu Y-C. Solvent-Dictated Lithium Sulfur Redox Reactions: An Operando UV–vis Spectroscopic Study. *J Phys Chem Lett.* 2016;7:1518–1525.
10. Cuisinier M, Hart C, Balasubramanian M, et al. Radical or Not Radical: Revisiting Lithium-Sulfur Electrochemistry in Nonaqueous Electrolytes. *Adv Energy Mater.* 2015;5:1401801.
11. Lutz L, Alves Dalla Corte D, Tang M, et al. Role of Electrolyte Anions in the Na–O_2 Battery: Implications for NaO_2 Solvation and the Stability of the Sodium Solid Electrolyte Interphase in Glyme Ethers. *Chem Mater.* 2017;29:6066–6075.
12. Kwabi DG, Tułodziecki M, Pour N, et al. Controlling Solution-Mediated Reaction Mechanisms of Oxygen Reduction Using Potential and Solvent for Aprotic Lithium–Oxygen Batteries. *J Phys Chem Lett.* 2016;7:1204–1212.
13. Yildirim H, Kinaci A, Chan MKY, et al. First-Principles Analysis of Defect Thermodynamics and Ion Transport in Inorganic SEI Compounds: LiF and NaF. *ACS Appl Mater Interfaces.* 2015;7:18985–18996.
14. Xu Y, Zhu Y, Liu Y, et al. Electrochemical Performance of Porous Carbon/Tin Composite Anodes for Sodium-Ion and Lithium-Ion Batteries. *Adv Energy Mater.* 2013;3:128–133.

15. Nayak PK, Yang L, Brehm W, et al. From Lithium-Ion to Sodium-Ion Batteries: Advantages, Challenges, and Surprises. *Angew Chem Int Ed Engl.* 2018;57:102–120.
16. Xu C, Yang Y, Wang H, et al. Electrolytes for Lithium- and Sodium-Metal Batteries. *Chem.* 2020;15:3584–3598.
17. Chen X, Yao N, Zeng B-S, et al. Ion–Solvent Chemistry in Lithium Battery Electrolytes: From Mono-Solvent to Multi-Solvent Complexes. *Fundam Res.* 2021;1:393–398.
18. Chen X, Shen X, Li B, et al. Ion–Solvent Complexes Promote Gas Evolution from Electrolytes on a Sodium Metal Anode. *Angew Chemie Int Ed.* 2018;57:734–737.
19. Tian Z, Zou Y, Liu G, et al. Electrolyte Solvation Structure Design for Sodium Ion Batteries. *Adv Sci.* 2022;9:2201207.
20. Peljo P, Girault HH. Electrochemical Potential Window of Battery Electrolytes: The HOMO–LUMO Misconception. *Energy Environ Sci.* 2018;11:2306–2309.
21. Nikiforidis G, van de Sanden MCM, Tsampas MN. High and Intermediate Temperature Sodium–Sulfur Batteries for Energy Storage: Development, Challenges and Perspectives. *RSC Adv.* 2019;9:5649–5673.
22. Wang Y, Zhou D, Palomares V, et al. Revitalising Sodium–Sulfur Batteries for Non-High-Temperature Operation: A Crucial Review. *Energy Environ Sci.* 2020;13:3848–3879.
23. Kummer JT, Weber N. Battery Having a Molten Alkali Metal Anode and a Molten Sulfur Cathode. *US Patent.* 1968:US3413150A.
24. Goodenough JB, Hong HYP, Kafalas JA. Fast Na^+-Ion Transport in Skeleton Structures. *Mater Res Bull.* 1976;11:203–220.
25. Oshima T, Kajita M, Okuno A. Development of Sodium-Sulfur Batteries. *Int J Appl Ceram Technol.* 2005;1:269–276.
26. Liu M. *Degradation of Sodium Beta"-Alumina Electrolyte in Contact with Sulfur/Sodium Polysulfide Melts* (Doctoral Thesis); 1986. LBL-21563. https://escholarship.org/uc/item/1rk0002z.
27. Andriollo M, Benato R, Dambone Sessa S, et al. Energy Intensive Electrochemical Storage in Italy: 34.8MW Sodium–Sulphur Secondary Cells. *J Energy Storage.* 2016;5:146–155.
28. Chang R, Minck R. Sodium-Sulfur Battery Flight Experiment Definition Study. *J Power Sources.* 1990;29:555–563.
29. Abraham KM, Rauh RD, Brummer SB. A Low Temperature Na-S Battery Incorporating A Soluble S Cathode. *Electrochim Acta.* 1978;23:501–507.
30. Fielder WL, Singer J. Solubility, Stability, and Electrochemical Studies of Sulfur-Sulfide. *NASA Tech Pap.* 1978;19780020681:1–40.
31. Park C, Ryu H, Kim K, et al. Discharge Properties of All-Solid Sodium–Sulfur Battery Using Poly (Ethylene Oxide) Electrolyte. *J Power Sources.* 2007;165:450–454.
32. Wang J, Yang J, Nuli Y, et al. Room Temperature Na/S Batteries with Sulfur Composite Cathode Materials. *Electrochem Commun.* 2007;9:31–34.
33. Wang Y-X, Zhang B, Lai W, et al. Room-Temperature Sodium-Sulfur Batteries: A Comprehensive Review on Research Progress and Cell Chemistry. *Adv Energy Mater.* 2017;7:1602829.
34. Bauer I, Kohl M, Althues H, et al. Shuttle Suppression in Room Temperature Sodium–Sulfur Batteries Using Ion Selective Polymer Membranes. *Chem Commun.* 2014;50:3208.
35. Adelhelm P, Hartmann P, Bender CL, et al. From Lithium to Sodium: Cell Chemistry of Room Temperature Sodium–Air and Sodium–Sulfur Batteries. *Beilstein J Nanotechnol.* 2015;6:1016–1055.
36. Yu X, Manthiram A. Room-Temperature Sodium–Sulfur Batteries with Liquid-Phase Sodium Polysulfide Catholytes and Binder-Free Multiwall Carbon Nanotube Fabric Electrodes. *J Phys Chem C.* 2014;118:22952–22959.
37. Ryu H, Kim T, Kim K, et al. Discharge Reaction Mechanism of Room-Temperature Sodium-Sulfur Battery with Tetra Ethylene Glycol Dimethyl Ether Liquid Electrolyte. *J Power Sources.* 2011;196:5186–5190.

38. Matios E, Wang H, Wang C, et al. Enabling Safe Sodium Metal Batteries by Solid Electrolyte Interphase Engineering: A Review. *Ind Eng Chem Res.* 2019;58: 9758–9780.
39. Monti D, Jónsson E, Boschin A, et al. Towards Standard Electrolytes for Sodium-Ion Batteries: Physical Properties, Ion Solvation and Ion-Pairing in Alkyl Carbonate Solvents. *Phys Chem Chem Phys.* 2020;22:22768–22777.
40. Li Y, Lu Y, Adelhelm P, et al. Intercalation Chemistry of Graphite: Alkali Metal Ions and Beyond. *Chem Soc Rev.* 2019;48:4655–4687.
41. Raguette L, Jorn R. Ion Solvation and Dynamics at Solid Electrolyte Interphases: A Long Way from Bulk? *J Phys Chem C.* 2018;122:3219–3232.
42. Chen X, Shen X, Hou T-Z, et al. Ion-Solvent Chemistry-Inspired Cation-Additive Strategy to Stabilize Electrolytes for Sodium-Metal Batteries. *Chem.* 2020;6:2242–2256.
43. Andreev M, de Pablo JJ, Chremos A, et al. Influence of Ion Solvation on the Properties of Electrolyte Solutions. *J Phys Chem B.* 2018;122:4029–4034.
44. Vineeth SK, Soni CB, Sun Y, et al. Implications of Na-Ion Solvation on Na Anode–Electrolyte Interphase. *Trends Chem.* 2022;4:48–59.
45. Tiwari S, Gupta AK, Gupta S, et al. Studies on the Interaction of Na^+ Ion with Binary Mixture of Carbonate-Ester Solvents: A Density Functional Theory Approach. *J Phys Conf Ser.* 2021;1849:012024.
46. Ding MS, Xu K, Jow TR. Liquid-Solid Phase Diagrams of Binary Carbonates for Lithium Batteries. *J Electrochem Soc.* 2000;147:1688.
47. Karatrantos A, Khan S, Ohba T, et al. The Effect of Different Organic Solvents on Sodium Ion Storage in Carbon Nanopores. *Phys Chem Chem Phys.* 2018;20: 6307–6315.
48. Shakourian-Fard M, Kamath G, Smith K, et al. Trends in Na-Ion Solvation with Alkyl-Carbonate Electrolytes for Sodium-Ion Batteries: Insights from First-Principles Calculations. *J Phys Chem C.* 2015;119:22747–22759.
49. Wahlers J, Fulfer KD, Harding DP, et al. Solvation Structure and Concentration in Glyme-Based Sodium Electrolytes: A Combined Spectroscopic and Computational Study. *J Phys Chem C.* 2016;120:17949–17959.
50. Liu H, Lai W, Lei Y, et al. Electrolytes/Interphases: Enabling Distinguishable Sulfur Redox Processes in Room-Temperature Sodium-Sulfur Batteries. *Adv Energy Mater.* 2022;2103304.
51. Sungjemmenla, Soni CB, Vineeth SK, et al. Exploration of the Unique Structural Chemistry of Sulfur Cathode for High-Energy Rechargeable Beyond-Li Batteries. *Adv Energy Sustain Res.* 2022;3:2100157.
52. Kumar V, Wang Y, Eng AYS, et al. A Biphasic Interphase Design Enabling High Performance in Room Temperature Sodium- Sulfur Batteries. *Cell Reports Phys Sci.* 2020;1:100044.
53. Liu Y, Li X, Sun Y, et al. Macro-Microporous Carbon with a Three-Dimensional Channel Skeleton Derived from Waste Sunflower Seed Shells for Sustainable Room-Temperature Sodium Sulfur Batteries. *J Alloys Compd.* 2021;853:157316.
54. Kumar V, Eng AYS, Wang Y, et al. An Artificial Metal-Alloy Interphase for High-Rate and Long-Life Sodium–Sulfur Batteries. *Energy Storage Mater.* 2020;29:1–8.
55. Fan X, Yue J, Han F, et al. High-Performance All-Solid-State Na–S Battery Enabled by Casting–Annealing Technology. *ACS Nano.* 2018;12:3360–3368.
56. Song S, Duong HM, Korsunsky AM, et al. A Na^+ Superionic Conductor for Room-Temperature Sodium Batteries. *Sci Rep.* 2016;6:32330.
57. Zhu T, Dong X, Liu Y, et al. An All-Solid-State Sodium – Sulfur Battery Using a Sulfur/Carbonized Polyacrylonitrile Composite Cathode. *ACS Appl Energy Mater.* 2019;2:5263–5271.
58. Ge Z, Li J, Liu J. Enhanced Electrochemical Performance of All-Solid-State Sodium-Sulfur Batteries by $PEO-NaCF_3SO_3$-MIL-53(Al) solid electrolyte. *Ionics (Kiel).* 2020;26:1787–1795.

59. Hwang TH, Jung DS, Kim J-S, et al. One-Dimensional Carbon–Sulfur Composite Fibers for Na–S Rechargeable Batteries Operating at Room Temperature. *Nano Lett.* 2013;13:4532–4538.
60. Xin S, Yin Y, Guo Y, et al. A High-Energy Room-Temperature Sodium-Sulfur Battery. *Adv Mater.* 2014;26:1261–1265.
61. Wang H, Wang C, Matios E, et al. Facile Stabilization of the Sodium Metal Anode with Additives: Unexpected Key Role of Sodium Polysulfide and Adverse Effect of Sodium Nitrate. *Angew Chemie.* 2018;130:7860–7863.
62. Feng J, Ci L, Xiong S. Biphenyl as Overcharge Protection Additive for Nonaqueous Sodium Batteries. *RSC Adv.* 2015;5:96649–96652.
63. Wu J, Liu J, Lu Z, et al. Non-Flammable Electrolyte for Dendrite-Free Sodium-Sulfur Battery. *Energy Storage Mater.* 2019;23:8–16.
64. Kumar D, Mishra K. A Brief Overview of Room Temperature Na-S Batteries Using Composite Sulfur Cathode. *Macromol Symp.* 2021;398:1900206.
65. Wang Y, Zhong W-H. Development of Electrolytes towards Achieving Safe and High-Performance Energy-Storage Devices: A Review. *ChemElectroChem.* 2015;2:22–36.
66. Lu D, Shao Y, Lozano T, et al. Failure Mechanism for Fast-Charged Lithium Metal Batteries with Liquid Electrolytes. *Adv Energy Mater.* 2015;5:1400993.
67. Liu T, Yang X, Nai J, et al. Recent Development of Na Metal Anodes: Interphase Engineering Chemistries Determine the Electrochemical Performance. *Chem Eng J.* 2021;409:127943.
68. Ren X, Zou L, Jiao S, et al. High-Concentration Ether Electrolytes for Stable High-Voltage Lithium Metal Batteries. *ACS Energy Lett.* 2019;4:896–902.
69. Yamada Y, Yamada A. Review—Superconcentrated Electrolytes for Lithium Batteries. *J Electrochem Soc.* 2015;162:A2406–A2423.
70. Henderson WA, Brooks NR. Crystals from Concentrated Glyme Mixtures. The Single-Crystal Structure of $LiClO_4$. *Inorg Chem.* 2003;42:4522–4524.
71. Henderson WA, Brooks NR, Brennessel WW, et al. Triglyme–Li^+ Cation Solvate Structures: Models for Amorphous Concentrated Liquid and Polymer Electrolytes (I). *Chem Mater.* 2003;15:4679–4684.
72. Brouillette D, Irish DE, Taylor NJ, et al. Stable Solvates in Solution of Lithium Bis(trifluoromethylsulfone)imide in Glymes and Other Aprotic Solvents: Phase Diagrams, Crystallography and Raman SpectroscopyElectronic Supplementary Information (ESI) Available: Crystallographic Data (Single Crystal). *Phys Chem Chem Phys.* 2002;4:6063–6071.
73. Okamoto Y, Tsuzuki S, Tatara R, et al. High Transference Number of Na Ion in Liquid-State Sulfolane Solvates of Sodium Bis(fluorosulfonyl)amide. *J Phys Chem C.* 2020;124:4459–4469.
74. Cao R, Mishra K, Li X, et al. Enabling Room Temperature Sodium Metal Batteries. *Nano Energy.* 2016;30:825–830.
75. Schafzahl L, Hanzu I, Wilkening M, et al. An Electrolyte for Reversible Cycling of Sodium Metal and Intercalation Compounds. *ChemSusChem.* 2017;10:401–408.
76. Lee J, Lee Y, Lee J, et al. Ultraconcentrated Sodium Bis(fluorosulfonyl)imide-Based Electrolytes for High-Performance Sodium Metal Batteries. *ACS Appl Mater Interfaces.* 2017;9:3723–3732.
77. Chen S, Zheng J, Mei D, et al. High-Voltage Lithium-Metal Batteries Enabled by Localized High-Concentration Electrolytes. *Adv Mater.* 2018;30:1706102.
78. Cao X, Jia H, Xu W, et al. Review—Localized High-Concentration Electrolytes for Lithium Batteries. *J Electrochem Soc.* 2021;168:010522.
79. Self J, Fong KD, Persson KA. Transport in Superconcentrated $LiPF_6$ and $LiBF_4$ / Propylene Carbonate Electrolytes. *ACS Energy Lett.* 2019;4:2843–2849.

80. Doi T, Shimizu Y, Hashinokuchi M, et al. Dilution of Highly Concentrated LiBF$_4$ / Propylene Carbonate Electrolyte Solution with Fluoroalkyl Ethers for 5-V LiNi$_{0.5}$Mn$_{1.5}$O$_4$ Positive Electrodes. *J Electrochem Soc.* 2017;164:A6412–A6416.
81. Xu K. Nonaqueous Liquid Electrolytes for Lithium-Based Rechargeable Batteries. *Chem Rev.* 2004;104:4303–4418.
82. Zheng J, Chen S, Zhao W, et al. Extremely Stable Sodium Metal Batteries Enabled by Localized High-Concentration Electrolytes. *ACS Energy Lett.* 2018;3:315–321.
83. Wang Y, Jiang R, Liu Y, et al. Enhanced Sodium Metal/Electrolyte Interface by a Localized High-Concentration Electrolyte for Sodium Metal Batteries: First-Principles Calculations and Experimental Studies. *ACS Appl Energy Mater.* 2021;4:7376–7384.
84. He J, Bhargav A, Shin W, et al. Stable Dendrite-Free Sodium–Sulfur Batteries Enabled by a Localized High-Concentration Electrolyte. *J Am Chem Soc.* 2021;143:20241–20248.
85. Qiao L, Judez X, Rojo T, et al. Review—Polymer Electrolytes for Sodium Batteries. *J Electrochem Soc.* 2020;167:070534.
86. Blatt MP, Hallinan DT. Polymer Blend Electrolytes for Batteries and Beyond. *Ind Eng Chem Res.* 2021;60:17303–17327.
87. Gadhave RV, Vineeth SK. Synthesis and Characterization of Starch Stabilized Polyvinyl Acetate-Acrylic Acid Copolymer-Based Wood Adhesive. *Polym Bull.* 2022. https://doi.org/10.1007/s00289-022-04558-8.
88. Vineeth SK, Gadhave RV, Gadekar PT. Glyoxal Cross-Linked Polyvinyl Alcohol-Microcrystalline Cellulose Blend as a Wood Adhesive with Enhanced Mechanical, Thermal and Performance Properties. *Mater Int.* 2020;2:0277–0285.
89. Vineeth SK, Gadhave RV, Gadekar PT. Polyvinyl Alcohol–Cellulose Blend Wood Adhesive Modified by Citric Acid and Its Effect on Physical, Thermal, Mechanical and Performance Properties. *Polym Bull.* 2022. https://doi.org/10.1007/s00289-022-04439-0.
90. Vineeth SK, Gadhave RV, Gadekar PT. Investigation of Crosslinking Ability of Sodium Metabisulphite with Polyvinyl Alcohol–Corn Starch Blend and Its Applicability as Wood Adhesive. *Indian Chem Eng.* 2022;64:197–207.
91. Gadhave RV, Vineeth SK, Dhawale PV, et al. Effect of Boric acid on Poly Vinyl Alcohol-Tannin Blend and Its Application as Water-Based Wood Adhesive. *Des Monomers Polym [Internet].* 2020;23:188–196.
92. Dhawale PV, Vineeth SK, Gadhave RV, et al. Tannin as a Renewable Raw Material for Adhesive Applications: A Review. *Mater Adv.* 2022;3:3365–3388.
93. Singh HK, Patil T, Vineeth SK, et al. Isolation of Microcrystalline Cellulose from Corn Stover with Emphasis on Its Constituents: Corn Cover and Corn Cob. *Mater Today Proc.* 2020;27:589–594.
94. Park C-W, Ahn J-H, Ryu H-S, et al. Room-Temperature Solid-State Sodium/Sulfur Battery. *Electrochem Solid-State Lett.* 2006;9:A123.
95. Kumar D, Suleman M, Hashmi SA. Studies on Poly(vinylidene fluoride-co-hexafluoropropylene) Based Gel Electrolyte Nanocomposite for Sodium–Sulfur Batteries. *Solid State Ionics.* 2011;202:45–53.
96. Kumar D. Effect of Organic Solvent Addition on Electrochemical Properties of Ionic Liquid Based Na$^+$ Conducting Gel Electrolytes. *Solid State Ionics.* 2018;318:65–70.
97. Kumar D, Kanchan DK. Dielectric and Electrochemical Studies on Carbonate Free Na-Ion Conducting Electrolytes for Sodium-Sulfur Batteries. *J Energy Storage.* 2019;22:44–49.
98. Kim J-S, Ahn H-J, Kim I-P, et al. The Short-Term Cycling Properties of Na/PVdF/S Battery at Ambient Temperature. *J Solid State Electrochem.* 2008;12:861–865.
99. Lim D-H, Agostini M, Ahn J-H, et al. An Electrospun Nanofiber Membrane as Gel-Based Electrolyte for Room-Temperature Sodium-Sulfur Batteries. *Energy Technol.* 2018;6:1214–1219.

100. Murugan S, Klostermann S V., Schützendübe P, et al. Stable Cycling of Room-Temperature Sodium-Sulfur Batteries Based on an In Situ Crosslinked Gel Polymer Electrolyte. *Adv Funct Mater.* 2022;2201191.
101. Zhou D, Chen Y, Li B, et al. A Stable Quasi-Solid-State Sodium–Sulfur Battery. *Angew Chemie Int Ed.* 2018;57:10168–10172.
102. Luo C, Shen T, Ji H, et al. Mechanically Robust Gel Polymer Electrolyte for an Ultrastable Sodium Metal Battery. *Small.* 2020;16:1906208.
103. Lei D, He YB, Huang H, et al. Cross-Linked Beta Alumina Nanowires with Compact Gel Polymer Electrolyte Coating for Ultra-Stable Sodium Metal Battery. *Nat Commun.* 2019;10:1–11.
104. Yang L, Jiang Y, Liang X, et al. Novel Sodium–Poly(tartaric acid)Borate-Based Single-Ion Conducting Polymer Electrolyte for Sodium–Metal Batteries. *ACS Appl Energy Mater.* 2020;3:10053–10060.
105. Zhang Z, Huang Y, Li C, et al. Metal–Organic Framework-Supported Poly(ethylene oxide) Composite Gel Polymer Electrolytes for High-Performance Lithium/Sodium Metal Batteries. *ACS Appl Mater Interfaces.* 2021;13:37262–37272.
106. Song S, Kotobuki M, Zheng F, et al. A Hybrid Polymer/Oxide/Ionic-liquid Solid Electrolyte for Na-Metal Batteries. *J Mater Chem A.* 2017;5:6424–6431.
107. Matios E, Wang H, Luo J, et al. Reactivity-Guided Formulation of Composite Solid Polymer Electrolytes for Superior Sodium Metal Batteries. *J Mater Chem A.* 2021;9:18632–18643.
108. Park M, Woo H, Heo J, et al. Thermoplastic Polyurethane Elastomer-Based Gel Polymer Electrolytes for Sodium-Metal Cells with Enhanced Cycling Performance. *ChemSusChem.* 2019;12:4645–4654.
109. Zheng Y, Pan Q, Clites M, et al. High-Capacity All-Solid-State Sodium Metal Battery with Hybrid Polymer Electrolytes. *Adv Energy Mater.* 2018;8:1801885.
110. Xu X, Lin K, Zhou D, et al. Quasi-Solid-State Dual-Ion Sodium Metal Batteries for Low-Cost Energy Storage. *Chem.* 2020;6:902–918.
111. Faktorovich-Simon E, Natan A, Peled E, et al. Oxygen Redox Processes in PEGDME-Based Electrolytes for the Na-Air Battery. *J Solid State Electrochem.* 2018;22:1015–1022.
112. Faraday M. I. Experimental Researches in Electricity—fifteenth Series. *Philos Trans R Soc London.* 1839;129:1–12.
113. He X, Zhu Y, Mo Y. Origin of Fast Ion Diffusion in Super-Ionic Conductors. *Nat Commun.* 2017;8:15893.
114. Sun F, Wang C, Osenberg M, et al. Clarifying the Electro-Chemo-Mechanical Coupling in $Li_{10}SnP_2S_{12}$ based All-Solid-State Batteries. *Adv Energy Mater.* 2022;12:2103714.
115. Chen Z, Zhang H, Xu H, et al. In Situ Generated Polymer Electrolyte Coating-Based Janus Interfaces for Long-Life LAGP-Based NMC811/Li Metal Batteries. *Chem Eng J.* 2022;433:133589.
116. Vineeth SK, Tebyetekerwa M, Liu H, et al. Progress in the Development of Solid-State Electrolytes for Reversible Room-Temperature Sodium–Sulfur Batteries. *Mater Adv.* 2022;3:6415–6440.
117. Nguyen H, Banerjee A, Wang X, et al. Single-Step Synthesis of Highly Conductive Na_3PS_4 Solid Electrolyte for Sodium All Solid-State Batteries. *J Power Sources.* 2019;435:126623.
118. Zhang Z, Shao Y, Lotsch B, et al. New Horizons for Inorganic Solid State Ion Conductors. *Energy Environ Sci.* 2018;11:1945–1976.
119. Berbano SS, Seo I, Bischoff CM, et al. Formation and Structure of $Na_2S + P_2S_5$ Amorphous Materials Prepared by Melt-Quenching and Mechanical Milling. *J Non Cryst Solids.* 2012;358:93–98.
120. Noi K, Hayashi A, Tatsumisago M. Structure and Properties of the Na_2S-P_2S_5 Glasses and Glass-Ceramics Prepared by Mechanical Milling. *J Power Sources.* 2014;269:260–265.

121. Famprikis T, Kudu ÖU, Dawson JA, et al. Under Pressure: Mechanochemical Effects on Structure and Ion Conduction in the Sodium-Ion Solid Electrolyte Na_3PS_4. *J Am Chem Soc.* 2020;142:18422–18436.
122. Krauskopf T, Culver SP, Zeier WG. Local Tetragonal Structure of the Cubic Superionic Conductor Na_3PS_4. *Inorg Chem.* 2018;57:4739–4744.
123. Chu IH, Kompella CS, Nguyen H, et al. Room-Temperature All-solid-state Rechargeable Sodium-ion Batteries with a Cl-doped Na_3PS_4 Superionic Conductor. *Sci Rep.* 2016;6:1–10.
124. Jansen M, Henseler U. Synthesis, Structure Determination, and Ionic Conductivity of Sodium Tetrathiophosphate. *J Solid State Chem.* 1992;99:110–119.
125. Yu C, Ganapathy S, De Klerk NJJ, et al. Na-Ion Dynamics in Tetragonal and Cubic Na_3PS_4, a Na-Ion Conductor for Solid State Na-Ion Batteries. *J Mater Chem A.* 2016;4:15095–15105.
126. Hayashi A, Noi K, Sakuda A, et al. Superionic Glass-Ceramic Electrolytes for Room-Temperature Rechargeable Sodium Batteries. *Nat Commun.* 2012;3:856.
127. Tanibata N, Noi K, Hayashi A, et al. Preparation and Characterization of Highly Sodium Ion Conducting Na_3PS_4–Na_4SiS_4 solid electrolytes. *RSC Adv.* 2014;4:17120–17123.
128. Zhu Z, Chu I-H, Deng Z, et al. Role of Na^+ Interstitials and Dopants in Enhancing the Na^+ Conductivity of the Cubic Na_3PS_4 Superionic Conductor. *Chem Mater.* 2015;27:8318–8325.
129. Rao RP, Chen H, Wong LL, et al. $Na_{3+x}M_xP_{1-x}S_4$ (M = Ge^{4+}, Ti^{4+}, Sn^{4+}) Enables High Rate All-Solid-State Na-ion Batteries $Na_{2+2d}Fe_{2-d}(SO_4)_3$|$Na_{3+x}M_xP_{1-x}S_4$|$Na_2Ti_3O_7$. *J Mater Chem A.* 2017;5:3377–3388.
130. Dawson JA, Canepa P, Clarke MJ, et al. Toward Understanding the Different Influences of Grain Boundaries on Ion Transport in Sulfide and Oxide Solid Electrolytes. *Chem Mater.* 2019;31:5296–5304.
131. Takeuchi S, Suzuki K, Hirayama M, et al. Sodium Superionic Conduction in Tetragonal Na_3PS_4. *J Solid State Chem.* 2018;265:353–358.
132. Tanibata N, Noi K, Hayashi A, et al. X-ray Crystal Structure Analysis of Sodium-Ion Conductivity in $94Na_3PS_4 \cdot 6Na_4SiS_4$ Glass-Ceramic Electrolytes. *ChemElectroChem.* 2014;1:1130–1132.
133. de Klerk NJJ, Wagemaker M. Diffusion Mechanism of the Sodium-Ion Solid Electrolyte Na_3PS_4 and Potential Improvements of Halogen Doping. *Chem Mater.* 2016;28:3122–3130.
134. Huang H, Wu H-H, Wang X, et al. Enhancing Sodium Ionic Conductivity in Tetragonal-Na_3PS_4 by Halogen Doping: A First Principles Investigation. *Phys Chem Chem Phys.* 2018;20:20525–20533.
135. Hayashi A, Noi K, Tanibata N, et al. High Sodium Ion Conductivity of Glass–Ceramic Electrolytes with Cubic Na_3PS_4. *J Power Sources.* 2014;258:420–423.
136. Yubuchi S, Hayashi A, Tatsumisago M. Sodium-ion Conducting Na 3 PS 4 Electrolyte Synthesized via a Liquid-phase Process Using N-Methylformamide. *Chem Lett.* 2015;44:884–886.
137. Nagata H, Chikusa Y. An All-solid-state Sodium–Sulfur Battery Operating at Room Temperature Using a High-sulfur-content Positive Composite Electrode. *Chem Lett.* 2014;43:1333–1334.
138. Yue J, Han F, Fan X, et al. High-Performance All-Inorganic Solid-State Sodium–Sulfur Battery. *ACS Nano.* 2017;11:4885–4891.
139. Wan H, Weng W, Han F, et al. Bio-inspired Nanoscaled Electronic/Ionic Conduction Networks for Room-Temperature All-Solid-State Sodium-Sulfur Battery. *Nano Today.* 2020;33:100860.
140. An T, Jia H, Peng L, et al. Material and Interfacial Modification toward a Stable Room-Temperature Solid-State Na–S Battery. *ACS Appl Mater Interfaces.* 2020;12:20563–20569.

141. Rajagopalan R, Zhang Z, Tang Y, et al. Understanding Crystal Structures, Ion Diffusion Mechanisms and Sodium Storage Behaviors of NASICON Materials. *Energy Storage Mater.* 2021;34:171–193.
142. Bohnke O, Ronchetti S, Mazza D. Conductivity Measurements on Nasicon and Nasicon-Modified Materials. *Solid State Ionics.* 1999;122:127–136.
143. Rao YB, Bharathi KK, Patro LN. Review on the Synthesis and Doping Strategies in Enhancing the Na Ion Conductivity of $Na_3Zr_2Si_2PO_{12}$ (NASICON) Based Solid Electrolytes. *Solid State Ionics.* 2021;366–367:115671.
144. Yu X, Manthiram A. Sodium-Sulfur Batteries with a Polymer-Coated NASICON-type Sodium-Ion Solid Electrolyte. *Matter.* 2019;1:439–451.
145. Lu L, Lu Y, Alonso JA, et al. A Monolithic Solid-State Sodium–Sulfur Battery with Al-Doped $Na_{3.4}Zr_2(Si_{0.8}P_{0.2}O_4)_3$ Electrolyte. *ACS Appl Mater Interfaces.* 2021;13:42927–42934.
146. Vignarooban K, Kushagra R, Elango A, et al. Current Trends and Future Challenges of Electrolytes for Sodium-Ion Batteries. *Int J Hydrogen Energy.* 2016;41:2829–2846.
147. Kummer JT. β-Alumina Electrolytes. *Prog Solid State Chem.* 1972;7:141–175.
148. Lee S-T, Lee D-H, Kim J-S, et al. Influence of Fe and Ti Addition on Properties of Na+-β/β″-Alumina Solid Electrolytes. *Met Mater Int.* 2017;23:246–253.
149. Wen Z, Gu Z, Xu X, et al. Research Activities in Shanghai Institute of Ceramics, Chinese Academy of Sciences on the Solid Electrolytes for Sodium Sulfur Batteries. *J Power Sources.* 2008;184:641–645.
150. Wei X, Cao Y, Lu L, et al. Synthesis and Characterization of Titanium Doped Sodium Beta″-Alumina. *J Alloys Compd.* 2011;509:6222–6226.
151. Barbosa JC, Gonçalves R, Costa CM, et al. Toward Sustainable Solid Polymer Electrolytes for Lithium-Ion Batteries. *ACS Omega.* 2022;7:14457–14464.
152. Ngai KS, Ramesh S, Ramesh K, et al. A Review of Polymer Electrolytes: Fundamental, Approaches and Applications. *Ionics (Kiel).* 2016;22:1259–1279.
153. Fergus JW. Ion Transport in Sodium Ion Conducting Solid Electrolytes. *Solid State Ionics.* 2012;227:102–112.
154. Gebert F, Knott J, Gorkin R, et al. Polymer Electrolytes for Sodium-Ion Batteries. *Energy Storage Mater.* 2021;36:10–30.
155. Lopez J, Mackanic DG, Cui Y, et al. Designing Polymers for Advanced Battery Chemistries. *Nat Rev Mater.* 2019;4:312–330.
156. Mallaiah Y, Jeedi VR, Swarnalatha R, et al. Impact of Polymer Blending on Ionic Conduction Mechanism and Dielectric Properties of Sodium Based PEO-PVdF Solid Polymer Electrolyte Systems. *J Phys Chem Solids.* 2021;155:110096.
157. Feng J, Wang L, Chen Y, et al. PEO Based Polymer-Ceramic Hybrid Solid Electrolytes: A Review. *Nano Converg.* 2021;8:2.
158. Jishnu NS, Vineeth SK, Das A, et al. Electrospun PVdF and PVdF-co-HFP-Based Blend Polymer Electrolytes for Lithium Ion Batteries. In: Balakrishnan NTM, Prasanth R, editors. *Electrospinning for Advanced Energy Storage Applications.* 1st ed. Singapore: Springer Singapore; 2021. pp. 201–234.
159. Prasanth R, Aravindan V, Srinivasan M. Novel Polymer Electrolyte Based on Cob-web Electrospun Multi component Polymer Blend of Polyacrylonitrile/poly(methyl methacrylate)/polystyrene for Lithium Ion Batteries – Preparation and Electrochemical Characterization. *J Power Sources.* 2012;202:299–307.
160. Kumar D, Hashmi SA. Ionic Liquid Based Sodium Ion Conducting Gel Polymer Electrolytes. *Solid State Ionics.* 2010;181:416–423.
161. Long L, Wang S, Xiao M, et al. Polymer Electrolytes for Lithium Polymer Batteries. *J Mater Chem A.* 2016;4:10038–10069.
162. Prasanth R, Shubha N, Hng HH, et al. Effect of Nano-clay on Ionic Conductivity and Electrochemical Properties of Poly(vinylidene fluoride) Based Nanocomposite Porous

Polymer Membranes and their Application as Polymer Electrolyte in Lithium Ion Batteries. *Eur Polym J.* 2013;49:307–318.

163. Shubha N, Prasanth R, Hoon HH, et al. Dual Phase Polymer Gel Electrolyte Based on Non-woven Poly(vinylidenefluoride-co-hexafluoropropylene)-layered Clay Nanocomposite Fibrous Membranes for Lithium Ion Batteries. *Mater Res Bull.* 2013;48:526–537.

164. Vincent CA. Ion Transport in Polymer Electrolytes. *Electrochim Acta.* 1995;40:2035–2040.

165. Kumar M, Srivastava N. Conductivity and Dielectric Investigation of NH4I-Doped Synthesized Polymer Electrolyte System. *Ionics (Kiel).* 2015;21:1301–1310.

166. Jinisha B, Anilkumar KM, Manoj M, et al. Poly (ethylene oxide) (PEO)-Based, Sodium Ion-Conducting, Solid Polymer Electrolyte Films, Dispersed with Al_2O_3 Filler, for Applications in Sodium Ion Cells. *Ionics (Kiel).* 2018;24:1675–1683.

167. Wang W, Alexandridis P. Composite Polymer Electrolytes: Nanoparticles Affect Structure and Properties. *Polymers (Basel).* 2016;8:387.

168. Yang J, Zhang H, Zhou Q, et al. Safety-Enhanced Polymer Electrolytes for Sodium Batteries: Recent Progress and Perspectives. *ACS Appl Mater Interfaces.* 2019;11:17109–17127.

169. Zhang Y, Lu W, Cong L, et al. Cross-linking Network based on Poly(ethylene oxide): Solid Polymer Electrolyte for Room Temperature Lithium Battery. *J Power Sources.* 2019;420:63–72.

170. Jinisha B, Anil Kumar KM, Manoj M, et al. Development of a Novel Type of Solid Polymer Electrolyte for Solid State Lithium Battery Applications Based on Lithium Enriched Poly (ethylene oxide) (PEO)/Poly (vinyl pyrrolidone) (PVP) Blend Polymer. *Electrochim Acta.* 2017;235:210–222.

171. Prasanth R, Shubha N, Hng HH, et al. Effect of Poly(ethylene oxide) on Ionic Conductivity and Electrochemical Properties of Poly(vinylidenefluoride) Based Polymer Gel Electrolytes Prepared by Electrospinning for Lithium Ion Batteries. *J Power Sources.* 2014;245:283–291.

172. Singh VK, Singh SK, Gupta H, et al. Electrochemical Investigations of $Na_{0.7}CoO_2$ Cathode with PEO-NaTFSI-BMIMTFSI Electrolyte as Promising Material for Na-Rechargeable Battery. *J Solid State Electrochem.* 2018;22:1909–1919.

173. Li L, Shan Y, Yang X. New Insights for Constructing Solid Polymer Electrolytes with Ideal Lithium-Ion Transfer Channels by Using Inorganic Filler. *Mater Today Commun.* 2021;26:101910.

174. Wang Q, Liu X, Cui Z, et al. A Fluorinated Polycarbonate Based All Solid State Polymer Electrolyte for Lithium Metal Batteries. *Electrochim Acta.* 2020;337:135843.

175. Tian L, Wang M, Xiong L, et al. Preparation and Performance of p(OPal-MMA)/PVDF Blend Polymer Membrane via Phase-Inversion Process for Lithium-Ion Batteries. *J Electroanal Chem.* 2019;839:264–273.

176. Raghavan P, Zhao X, Shin C, et al. Preparation and Electrochemical Characterization of Polymer Electrolytes Based on Electrospun Poly(vinylidene fluoride-co-hexafluoropropylene)/Polyacrylonitrile Blend/Composite Membranes for Lithium Batteries. *J Power Sources.* 2010;195:6088–6094.

177. Raghavan P, Zhao X, Kim J-K, et al. Ionic Conductivity and Electrochemical Properties of Nanocomposite Polymer Electrolytes Based on Electrospun Poly(vinylidene fluoride-co-hexafluoropropylene) with Nano-sized Ceramic Fillers. *Electrochim Acta.* 2008;54:228–234.

178. Vijaya Kumar Saroja AP, Rajamani A, Muthusamy K, et al. Repelling Polysulfides Using White Graphite Introduced Polymer Membrane as a Shielding Layer in Ambient Temperature Sodium Sulfur Battery. *Adv Mater Interfaces.* 2019;6:1901497.

179. Sungjemmenla, Vineeth SK, Soni CB, et al. Understanding the Cathode–Electrolyte Interphase in Lithium-Ion Batteries. *Energy Technol.* 2022;10:2200421.
180. Sun B, Xiong P, Maitra U, et al. Design Strategies to Enable the Efficient Use of Sodium Metal Anodes in High-Energy Batteries. *Adv Mater.* 2020;32:1903891.
181. Ma LA, Naylor AJ, Nyholm L, et al. Strategies for Mitigating Dissolution of Solid Electrolyte Interphases in Sodium-Ion Batteries. *Angew Chemie Int Ed.* 2021;60:4855–4863.
182. Sun H, Zhu G, Xu X, et al. A Safe and Non-flammable Sodium Metal Battery Based on an Ionic Liquid Electrolyte. *Nat Commun.* 2019;10:3302.
183. Cheng X-B, Zhao M-Q, Chen C, et al. Nanodiamonds Suppress the Growth of Lithium Dendrites. *Nat Commun.* 2017;8:336.
184. Kreissl JJA, Langsdorf D, Tkachenko BA, et al. Incorporating Diamondoids as Electrolyte Additive in the Sodium Metal Anode to Mitigate Dendrite Growth. *ChemSusChem.* 2020;13:2661–2670.
185. Zheng X, Fu H, Hu C, et al. Toward a Stable Sodium Metal Anode in Carbonate Electrolyte: A Compact, Inorganic Alloy Interface. *J Phys Chem Lett.* 2019;10:707–714.
186. Wenzel S, Metelmann H, Raiß C, et al. Thermodynamics and Cell Chemistry of Room Temperature Sodium/Sulfur Cells with Liquid and Liquid/Solid Electrolyte. *J Power Sources.* 2013;243:758–765.
187. Ren YX, Jiang HR, Zhao TS, et al. Remedies of Capacity Fading in Room-Temperature Sodium-Sulfur Batteries. *J Power Sources.* 2018;396:304–313.
188. Soni CB, Kumar V, Seh ZW. Guiding Uniform Sodium Deposition through Host Modification for Sodium Metal Batteries. *Batter Supercaps.* 2022;5:e202100207:1–8.
189. Soni CB, Arya N, Sungjemmenla, et al. Microarchitectures of Carbon Nanotubes for Reversible Na Plating/Stripping toward the Development of Room-Temperature Na–S Batteries. *Energy Technol.* 2022;10:2200742.
190. Liu S, Liu D, Wang S, et al. Understanding the Cathode Electrolyte Interface Formation in Aqueous Electrolyte by Scanning Electrochemical Microscopy. *J Mater Chem A.* 2019;7:12993–12996.
191. Bai P, Han X, He Y, et al. Solid Electrolyte Interphase Manipulation towards Highly Stable Hard Carbon Anodes for Sodium Ion Batteries. *Energy Storage Mater.* 2020;25:324–333.
192. Michan AL, Parimalam BS, Leskes M, et al. Fluoroethylene Carbonate and Vinylene Carbonate Reduction: Understanding Lithium-Ion Battery Electrolyte Additives and Solid Electrolyte Interphase Formation. *Chem Mater.* 2016;28:8149–8159.
193. Eshetu GG, Diemant T, Hekmatfar M, et al. Impact of the Electrolyte Salt Anion on the Solid Electrolyte Interphase Formation in Sodium Ion Batteries. *Nano Energy.* 2019;55:327–340.
194. Eshetu GG, Elia GA, Armand M, et al. Electrolytes and Interphases in Sodium-Based Rechargeable Batteries: Recent Advances and Perspectives. *Adv Energy Mater.* 2020;10:2000093.
195. Okoshi M, Yamada Y, Yamada A, et al. Theoretical Analysis on De-Solvation of Lithium, Sodium, and Magnesium Cations to Organic Electrolyte Solvents. *J Electrochem Soc.* 2013;160:A2160–A2165.
196. Liu Q, Wu F, Mu D, et al. A Theoretical Study on Na^+ Solvation in Carbonate Ester and Ether Solvents for Sodium-Ion Batteries. *Phys Chem Chem Phys.* 2020;22:2164–2175.
197. Eng AYS, Soni CB, Lum Y, et al. Theory-guided Experimental Design in Battery Materials Research. *Sci Adv.* 2022;8. https://doi.org/10.1126/sciadv.abm2422.
198. Wheatle BK, Fuentes EF, Lynd NA, et al. Design of Polymer Blend Electrolytes through a Machine Learning Approach. *Macromolecules.* 2020;53:9449–9459.

5 Analytical techniques to probe room-temperature sodium-sulfur batteries

Chhail Bihari Soni, Sungjemmenla, and Vipin Kumar

5.1 INTRODUCTION

RT Na-S battery performance degradation is often associated with the failure of the solid electrolyte interphase (SEI) and polysulfide dissolution [1,2]. The irreversible volume change, gas evolution, and unwarranted dendrite formation lead to SEI failure [3], while polysulfide dissolution occurs due to the inability of the sulfur host to entrap the polysulfide [4]. The polysulfide dissolution could also be associated with the electrolyte and the chemical functionality of the separator [5]. The remedial strategies have been developed over the years; however, understanding their fundamental reasons remained a mystery. Attempts have been made to unveil the structural and chemical changes that occur during charge and discharge reactions of RT Na-S batteries [6]. Post-mortem or ex situ analysis through Scanning electron microscopy [7], Transmission electron microscopy [8,9], Raman spectroscopy [10], or X-ray photoelectron spectroscopy [11], ultraviolet-visible (UV-Vis) spectroscopy has often been conducted to monitor changes in the state of the electrodes. Though the ex situ characterization tools offer a deeper understanding of the electrode materials, obtaining real-time information about those changes remained difficult. Since the vital information of a functional cell, i.e., SEI growth, dendrite growth, gas evolution, electrolyte decomposition, the release of polysulfide, dissolution of polysulfide, polysulfide cross-over, etc., is nearly impossible to obtain through ex situ or post-mortem analysis. In the past few years, RT Na-S batteries have witnessed a significant rise in the number of in situ/operando characterization and analysis instruments [3,12,13]. The main difference between in situ/operando and ex situ techniques is the measurement environment or conditions. The ex situ analyses are often performed in an environment that is different from the cell's actual environment. While in situ/operando analyses are performed during cell's operation, and therefore, the cell's conditions remain unchanged. In situ/operando technique is superior to ex situ examination in the following aspects: (i) air-induced side reactions can be avoided entirely without disassembly, and analytical results will be more trustworthy; (ii) a continuous test can gather a variety of useful data, including metastable data, and aid in exploring how the electrodes' structure vary with reaction time; (iii) in situ approaches can examine the electrode's thermodynamic conditions and advance understanding of the dynamic process. In a nutshell, in situ characterization methods could be promising to offer a thorough understanding of the intricate reaction mechanism that is difficult to see during metal deposition and dissolution cycles. Therefore, a new set of characterization techniques needs to be developed to establish a real-time structure-property relationship for each component of RT Na-S batteries.

5.2 OVERVIEW OF THE ROUTINE TECHNIQUES TO PROBE SODIUM METAL ANODE

Generally, the analytical tools are of two broad types. The physical tools, i.e., the one which gives information about structural changes. However, the tools that provide information about chemical changes are called chemical tools. Examining sodium's physical and chemical information may provide clues about the underlying electrochemical processes and reaction mechanisms. For example, the roughness and particle size of the sodium's surface indicates the nonuniformity of the electrochemical reactions at a macroscopic level and uneven distribution of the local electric field. Microscopic tools, for instance, optical microscopy (OM) and electron microscopy, have often been employed to monitor the growth of dendrites and the uniformity of sodium deposition. Electron microscopy offers two-dimensional information from a few micrometers to a nanometer range. While OM can provide a larger scale of information, it cannot resolve the fine features. Transmission electron microscopy offers three-dimensional details on the surface. The sodium metal loses its integrity under a high-energy electron beam. Cryo-transmission electron microscopy, developed for biological samples, has recently been used to view the surface morphology of sodium at an atomic level [4]. Due to low-temperature conditions (at liquid nitrogen temperature), the sodium metal could withstand the high-energy electron beam. Though Cryo-TEM could provide atomistic information, the sample preparation is tedious and requires good technical skills to perform testing the analysis [14].

Unveiling the chemical information of the sodium is a little intricate and requires sophisticated tools. X-ray photoelectron spectroscopy (XPS) [7], also called electron spectroscopy for chemical analysis, is a tool that can capture the chemical information of the surface and below the surface (i.e., depth profiling). The chemical composition of the surface, i.e., SEI, can be extracted through high-resolution XPS. In addition, the information about the thickness of SEI can be extracted through depth profiling [15]. The interaction between the X-ray beam and the organic constituent of the SEI poses reliability concerns. The surface of a cycled sodium metal anode comprises inorganic and organic components. The former causes surface charging due to their inherent insulating nature. Moreover, it requires an entirely inert atmosphere to transfer the sample from the electrochemical to the vacuum environment. Since XPS requires a high vacuum environment, a trace amount of electrolyte may affect the accuracy of the measurement.

X-ray diffraction (XRD) has also been used to evaluate the surface information of sodium metal anode [15]. To perform XRD, a specialized sample holder is employed

Analytical techniques to probe RT Na-S batteries 117

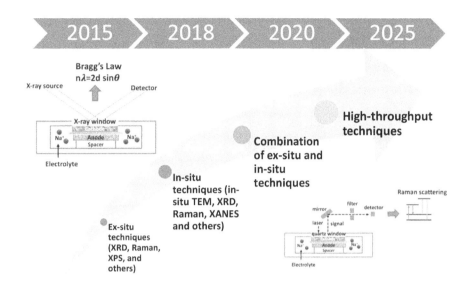

FIGURE 5.1 Timeline of the development of characterization techniques for sulfur cathode from 2015 to 2025.

to avoid the interaction of air or moisture with the cycled sodium. The formation of distinct chemical phases can be unveiled through XRD. The analysis area is limited to the X-ray beam size; highly localized surface information could be obtained. However, due to a relatively smaller proportion of distinct chemical phases, the background signal of sodium often dominates.

From the discussion, it is apparent that the routine characterization techniques are unable to probe the true nature of the sodium surface. New and innovative high-throughput characterization tools that can capture physical and chemical information simultaneously need to be developed to advance the understanding of the surface chemistry of sodium metal. Figure 5.1 schematically illustrates the progress made in developing the characterization tools to examine the nature of sodium dendrites.

5.3 OVERVIEW OF THE ROUTINE TECHNIQUES TO PROBE SULFUR CATHODE

Theoretically, the mechanistic route of transition from elemental sulfur (S_8) to short-chain polysulfides (Na_2S) has been clearly understood over the years [4,16,17]. However, understanding the underlying mechanisms and their correlation with their material behavior and characteristics remains questionable [18]. Many studies have been done on various characterization techniques for sulfur cathodes: ex situ and in situ [19–21]. Advanced characterization tools are of profound importance to understanding (i) the limiting factors that curb the capacity increase in the battery system, (ii) how the physical-chemical state of sulfur composites influences the transition behavior of sulfur moieties, and (iii) the state of the electrode after a certain number of charge-discharge cycles [21,22]. Notably, the electrochemical processes of dissolution

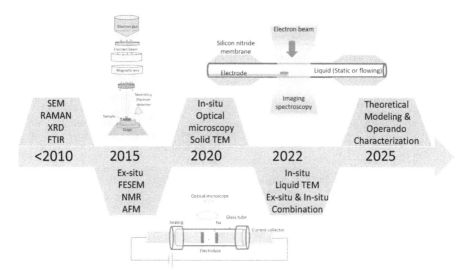

FIGURE 5.2 Timeline of the development of characterization techniques for Na metal anode from 2010 to 2025+.

or nucleation will occur over a range spanning from the micro-to-nanometer scale. Considering the sensitivity of the transition reduction moieties, various tools have been put forward to reflect and study the state of the battery. To retrieve information on the electrochemical reactions or monitor the responses during charge-discharge phenomena, various tools such as ex- and in situ Raman spectroscopy, ex-and in situ XRD, XPS, UV-vis, X-ray absorption near edge structure (XANES), and others. Figure 5.2 schematically illustrates the progress made in developing the characterization tools to examine the nature of the sodium polysulfide.

5.3.1 X-RAY DIFFRACTION

XRD is a non-destructive technique to probe the changes in the chemical phase of the cathode upon discharge and charge reactions. It works on the principle of diffraction of X-rays; the diffracted X-rays can interfere constructively or destructively depending on the angle, wavelength of the X-rays, and position of atoms. With the change in the position of the sample with respect to incident X-rays, the intensity of X-rays gets changed, leading to forming the patterns. These peaks represent the diffraction pattern of the crystal and are used to determine the crystal structure of a material.

To study the characteristics of the sulfur cathode, XRD has been conducted in both ex situ and in situ conditions. Since the environmental and test requirements for ex situ XRD critically affect the accuracy of the results, in situ XRD has often been employed to gauge chemical changes in the sulfur cathode. A typical sulfur discharge curve comprises four different phase transitions, i.e., liquid to a liquid state (Na_2S_8–Na_2S_6), liquid to a quasi-liquid-state (Na_2S_6–Na_2S_4), quasi-liquid to quasi-solid state (Na_2S_4–Na_2S_2) and quasi-solid to solid state (Na_2S_2–Na_2S). The formation of these polysulfides is susceptible to cell conditions and voltage [23,24]. For instance, Xiao et al. [25] studied the XRD patterns of the sulfur electrode during the discharge process. At around 1.6 V, a new peak was observed at around 30.8°, which can be attributed to the formation of Na_2S_4. On discharging further, another peak appeared at 23.4°, which is indexed to Na_2S formation. Fully charging it to 2.8 V, S_8 peaks fully appeared, confirming the full reversibility and reduction of short-chain sodium polysulfides.

In another set of in situ experiments, the cathode was discharged to 2.2 V, and the diffraction peaks positioned at 20.2° and 12.9° confirmed the presence of higher-order long-chain polysulfides (Na_2S_8 to Na_2S_6). When further discharged to 1.6 V, characteristic peaks at 11.9°, 20.1°, and 20.6° suggest the formation of Na_2S_5. On continuous discharge to 1.0 V, a new XRD peak appeared at 11.6°, was corresponds to Na_2S_4. Furthermore, this peak faded to give rise to a new peak for Na_2S_2 at 12.1°. Finally, discharging it to 0.7 V, a characteristic peak at 10.6°, attributed to Na_2S. When it was charged back to 2.8 V, the peaks corresponding to Na_2S mostly disappeared and were oxidized to higher-order polysulfides. Therefore, from in situ XRD, it is possible to monitor real-time changes in the chemical and electrochemical states of the sulfur cathode.

5.3.2 UV-VIS SPECTROSCOPY

It is a technique to study the electronic properties of materials through the absorption of light in the UV-visible region of the electromagnetic spectrum. The strength of absorption critically depends on the electronic, chemical, and physical properties of the sample. A relationship between absorption and wavelength of light is used to extract electronic and chemical information about the materials [26].

UV-Vis spectroscopy has been utilized to differentiate among various types of polysulfides. Based on the nature of absorption, the long and short-chain polysulfides can be distinguished [22]. Relating to a wavelength range of 800–200 nm, the UV visible range covers an energy spectrum range of 1.5–6.2 eV [27]. Considering the series of absorption in the spectrum, the characteristic absorption peaks can be detected for sulfur and its various discharge products [28]. Guo and co-workers contributed RT-Na/S battery technology without polysulfide dissolution based on a slit ultra-microporous carbon host [29]. Their hypothesis was supported by UV-vis spectroscopy. The findings demonstrated negligible changes of Na_2S in the UV-vis signal at various discharge states due to its limited solubility in the electrolyte system. Note that during the complete discharge process, Na_2S was found to be the only final discharge product based on the first-order derivative curves. A one-step reduction mechanism could thus be confirmed due to the hindrance in forming soluble and insoluble sodium polysulfides during the discharge reaction.

5.3.3 RAMAN SPECTROSCOPY

Raman spectroscopy works on the principle of scattering. When a beam of monochromatic light interacts with the matter, it gets scattered elastically and inelastically. The inelastic scattering, which carries the information about the loss or gain in energy, is the basis of the Raman spectrum. The Raman spectrum can identify the chemical composition, including functional groups and molecular bonds, by comparing the characteristic Raman frequencies of different molecules.

Raman spectroscopy has been used to detect the qualitative and semi-quantitative analysis of long-chain polysulfides dissolved in a conventional electrolyte [10]. The functional groups in the composite within the molecular structure allow excitation through vibrational modes, resulting in inelastic scattering of monochromatic light [30–33]. Raman will enable us to detect these signals [34]. Ex situ Raman has been used for room-temperature sodium-sulfur (RT-Na/S) batteries to investigate the structural changes and compositional diversions with respect to polysulfide transitions after the charge or discharge process [22]. Raman analysis was conducted to study the transition peaks from sulfur to polysulfides. During the static state, sulfur showed a characteristic peak at 482 cm^{-1}, which gradually disappeared upon discharging to 1.6 V. Similarly, Raman signals were observed for Na_2S_4 and Na_2S upon discharging, confirming the reduction of sulfur moieties [25].

However, in situ Raman has been used to detect the qualitative and semi-quantitative analysis of long-chain polysulfides dissolved in a conventional electrolyte [22]. For a fresh cell, Raman characteristic peaks for S_8 appear at 259 and 456 cm^{-1}. When discharged to a specific voltage (2.3 V), the peaks at S_8 become weaker, and a set of new peaks emerge at 374 and 473 cm^{-1}, which corresponds to Na_2S_6 and Na_2S_4, respectively. When discharged further (1.2 V), a peak at 428 cm^{-1} indicates the formation of Na_2S, and its intensity increases as the voltage is further reduced to 0.5 V. S_8 peaks, however, do not fully diminish, suggesting the presence of some S_8 throughout the process which may be due to its insulating nature and poor reactivity. It has been considered a powerful technique to probe and study real-time phase transitions of the sulfur cathode.

5.3.4 X-RAY ABSORPTION NEAR EDGE STRUCTURE

XANES spectroscopy works on the principle of X-rays absorption by the core level of an atom in the range of 50 eV. Since absorption is sensitive to the nature of the materials, the near-edge structure can reveal information about the oxidation state, coordination number, and chemical environment of the absorbing atoms. The XANES spectrum can also be used to identify the chemical composition of the sample [35,36].

XANES offers an advantage over the other spectroscopic tools for analyzing the electronic structure of sulfur cathodes to identify different chemical states of the sulfur cathode during the charge/discharge process [37]. This information can be used to optimize the composition and structure of the cathode to improve the cell performance [22,38]. Liu et al. [39] confirmed the effect of multisulfiphilic sites in electrocatalyzing sulfur cathodes via XANES. Before sulfur infusion, the Co K-edge of XANES of a core-shell matrix shifted to a higher energy state of 7730 eV, confirming the change in the oxidation state of cobalt from 0 to +2. After the infusion of sulfur, the Zn region of XANES shows a relatively strong absorption edge at 9668 eV, attributed to the change in the oxidation of Zn from 0 to +2. The findings show the chemisorption of the multisulfiphilic ZnS and CoS_2 is high toward sodium polysulfides. The formation of Zn-S and Co=S bonds occurs upon chemisorption. By providing insights into the composition and chemical properties of the cathode material, XANES can help researchers develop strategies to improve the performance and stability of sulfur-based batteries.

5.3.5 X-RAY PHOTOELECTRON SPECTROSCOPY

XPS works on the principle of photoelectric effect. The kinetic energy of the photoelectron gives information about the chemical bonds, oxidation states, and surface chemistry of the material. It can also be used to determine the near-surface (within a few nm) chemical compositions of the materials.

XPS is a powerful technique that can provide valuable information about the chemical and electronic properties of the sodium metal anode and sulfur cathodes [40]. Note that it is essential for understanding their performance and improving their design. Sulfur undergoes complex redox reactions during charge and discharge cycles, which can form various polysulfide species [41]. XPS can be used to identify and quantify these species and to study their spatial distribution within the cathode [42]. Understanding the formation and behavior of polysulfides is crucial for improving the capacity retention and cycling stability of sulfur cathodes. Chen and co-workers studied the untying of the thioether bond structure of carbon-sulfur for a RT Na-S battery system [43]. They have demonstrated the confinement of sulfur atoms in a thioether bond structure based on the deconvoluted low binding energy peaks of sulfur S 2p. In addition, low intense peaks with high binding energy peaks could also be confirmed, attributing to a lower content of $C-S(O)_2-C$ bonds. Furthermore, XPS can be used to study the surface chemistry and stability of sulfur cathodes. For instance, XPS can be used to investigate the formation and evolution of the cathode electrolyte interphase (CEI) layer on the surface of the cathode. The CEI layer is formed because of the reaction between the cathode and the electrolyte and plays a crucial role in the performance and safety of the battery. However, the formation of the CEI is highly debatable.

The chemical and physical complexity of the SEI made it challenging to determine its true nature accurately; however, due to the surface sensitivity of the XPS, it can determine the elemental composition of the SEI. Since the binding energy is sensitive to the chemical environment, any change in the binding energy gives information about the nature of the compound. For example, the binding energy of F is about 684–685.0 eV in the environment of metal atoms. The binding energy of F is observed to shift to 688–689 eV in the organic matter environment, suggesting the formation of carbon fluoride. Generally, the top surface of SEI comprises organic (RCH_2ONa) and inorganic matter (NaF, $NaCO_3$, Na_2O, $LiOH$). It is worth highlighting that the relative concentration of each organic or inorganic phase is subject to the nature of the electrolyte salt. For example, a fluoride-rich electrolyte (NaOTf or NaFSI or NaTFSI in diglyme) favors the formation of an inorganic phase. In contrast, a chloride-rich electrolyte (e.g., $NaClO_4$ in diglyme) leads to forming an organic-rich phase. Likewise, the electrolyte solvent affects the composition of inorganic and organic phases in the SEI. The relative binding energy position of each phase offers information about the nature and chemical composition of the SEI. Though XPS is a potential tool to approximately predict the chemical composition of the SEI, a few issues are yet to be resolved to rely on XPS predictions completely.

The coexistence of multiple chemical phases increases the complexity, as they may share a similar binding energy range. For example, the inorganic phase of the

SEI may comprise Na_2O, Na_2CO_3, and NaOH, which carry an identical ratio of Na-O. These issues may lead to incorrect or false predictions about the chemical nature of the SEI.

5.4 IN SITU/OPERANDO TECHNIQUES FOR SODIUM-SULFUR BATTERIES

The routine characterization techniques provide average and minimum information about the chemical and physical changes that occur during reactions, and they often do not meet the requirement of rigorous exploration. The vital information, for instance, nucleation, growth, and phase change, is difficult to extract from conventional characterization tools.

In recent years, effects have been made to characterize the anode or cathode materials through in situ/operando techniques. Following in situ/operando techniques have frequently been employed to extract ate chemical and physical information on the surface of the electrode.

5.4.1 IN SITU OM

In situ OM technique works on the principle of contrast (change in amplitude of brightness), i.e., change in intensity or color due to change in the cell conditions. The voltage or current can influence the cell conditions. It consists of an objective lens to create a magnified image of the object, and an eyepiece is used to magnify the object further. The contrast change in an electrochemical cell is mainly caused by a difference in the surface morphology or electrolyte conditions [44,45].

Examining the physical condition of the electrode and electrolytes through optical microscopy is appropriate due to its quick operation and its ability to monitor changes at a relatively larger scale. In addition, the direct observation of nucleation and growth of sodium dendrites can be made, and the relation with morphological features and sodium deposition/dissolution overpotential can be established. The information about the volume change can also be extracted through optical microscopy. Though it is difficult to get information about the SEI, movement/cracking of the surface that leads to dendrite growth can be observed under optical microscopy. This technique allows to get vital information about the dead sodium and its effect on the sodium deposition/dissolution overpotential.

The optical cell and its design are critical in extracting the information. Generally, two types of design are discussed in the literature, i.e., a flat cell and a capillary design. The capillary-type cell design allows only the cross-sectional information [46], while the top surface of the electrode can be observed in the flat cell design. The capillary cell design has often been used due to its ease of preparation and availability. A capillary tube of transparent glass combined with metallic working and counter/auxiliary electrodes makes up the test setup. Currently, in situ microscopy studies are performed to comprehend the deposition/dissolution of sodium over the Na metal anode [44]. Investigations revealed a strong correlation between Na granular deposition and deposition overpotential and tried to develop growth models for sodium dendrites.

5.4.2 IN SITU ATOMIC FORCE MICROSCOPY (AFM)

AFM works on the principle of Hooks law [47], where the force is measured between a sharp tip and the surface at a very short distance. In situ AFM is an arrangement with a three-electrode electrochemical cell inside the AFM environment. The lateral design of AFM generates a 2D surface or 3D morphological information of the working electrode. A ring-shaped counter electrode with a fine reference electrode is used to fabricate liquid electrochemical cells for in situ AFM imaging and testing. The local information of the working electrode can be obtained by scanning the AFM tip close to the electrode surface under varying potentials.

In situ AFM is widely used to explore the interface and surface features of the sodium metal anode with a resolution down to a few-angstrom [48]. Besides imaging, it can be used to analyze the mechanical properties of the SEI and the strength of dendrites [49].

5.4.3 IN SITU ELECTRON-MICROSCOPY TECHNIQUES

Electron microscopy techniques work on the principle of either scanning the electron beam over the surface of the sample or transmitting the electron beam through the sample [50]. The signals, i.e., secondary electrons or X-rays, are generated by electron and matter interaction. An electronic conductive region appears brighter, while less conductive or non-conductive regions appear darker under an electron microscope. Since the SEI is relatively less conductive than the sodium metal itself, the difference in the contrast forms the basis for the assessment of the SEI. Since most electron microscopy techniques require a high vacuum to function, limiting the direct use of liquid electrochemical cells. Moreover, most organic electrolytes cannot generate secondary electrons or diffract the electron beam, making direct visualization of dendritic growth.

In situ SEM or FESEM analysis of the dendrites or SEI is challenging; therefore, most of the SEM or FESEM is conducted under ex situ mode. However, specialized chambers are designed to ensure the chemical environment of the cell remains unchanged while dismantling the cell. Though employing the solid electrolytes for sodium metal or RT Na-S batteries is a tricky and complicated process, in situ SEM or FESEM analysis with solid electrolytes is yet to be realized.

Due to the high energy of the electron beam, the growth of dendrites and the SEI can be visualized under in situ TEM. For alkali metal anode batteries, the testing is generally performed in a symmetric cell design with solid electrolytes, i.e., MO_2 (M = Li or Na) [51,52]. As a result, the electrochemical environment does not cause degradation in the vacuum quality of the test chamber. The intrinsic structural and chemical changes at the electrode surface can be assessed, and a high-resolution image of the surface can be developed [53]. The dynamic sodium deposition morphology over a flat and curved surface can be captured through an in situ TEM.

5.5 FUTURE PROSPECTS

The sodium metal or RT Na-S battery research is in its infancy stage. Impressive advancements have been made to understand the growth of sodium dendrite, SEI, polysulfide formation, and shuttling. The cell performance critically depends on the physical and chemical properties of the electrodes and electrolyte and their mutual interaction. In the ex situ techniques, where the test environment is different from the cell environment, a convincing model to govern the dendrite growth, SEI growth, polysulfide formation, and dissolution could not be realized. Decent progress in studying SEI and dendrite growth has been made in the past few years; there is a dearth of literature on the formation and dissolution of sodium polysulfide.

In situ techniques offer a broader spectrum to analyze the chemical and physical properties of electrode materials. The interaction of the materials and the source is a matter of concern.

A hybrid approach with complementary ex situ and in situ techniques is needed for RT Na-S battery research.

REFERENCES

1. Eng AYS, Soni CB, Lum Y, et al. Theory-Guided Experimental Design in Battery Materials Research. *Sci Adv.* 2022;8. https://doi.org/10.1126/sciadv.abm2422.
2. Wang Y, Zhou D, Palomares V, et al. Revitalising Sodium-Sulfur Batteries for Non-High-Temperature Operation: A Crucial Review. *Energy Environ Sci.* 2020;13:3848–3879.
3. Soni CB, Sungjemmenla, Vineeth SK, et al. Challenges in Regulating Interfacial-Chemistry of the Sodium-Metal Anode for Room-Temperature Sodium-Sulfur Batteries. *Energy Storage.* 2022;4:e264.
4. Sungjemmenla, Soni CB, Vineeth SK, et al. Unveiling the Physiochemical Aspects of the Matrix in Improving Sulfur-Loading for Room-Temperature Sodium–Sulfur Batteries. *Mater Adv.* 2021;2:4165–4189.
5. Vineeth SK, Soni CB, Sun Y, et al. Implications of Na-Ion Solvation on Na Anode–Electrolyte Interphase. *Trends Chem.* 2022;4:48–59.
6. Vineeth SK, Tebyetekerwa M, Liu H, et al. Progress in the Development of Solid-State Electrolytes for Reversible Room-Temperature Sodium–Sulfur Batteries. *Mater Adv.* 2022;3:6415–6440.
7. Soni CB, Arya N, Sungjemmenla, et al. Microarchitectures of Carbon Nanotubes for Reversible Na Plating/Stripping Toward the Development of Room-Temperature Na–S Batteries. *Energy Technol.* 2022;10:2200742.
8. Du W, Xu Q, Zhan R, et al. Synthesis of Hollow Porous Carbon Microspheres and their Application to Room-Temperature Na-S Batteries. *Mater Lett.* 2018;221:66–69.
9. Jishnu NS, Vineeth SK, Das A, et al. Electrospun PVdF and PVdF-co-HFP-Based Blend Polymer Electrolytes for Lithium Ion Batteries. In: Balakrishnan NTM, Prasanth R, editors. *Electrospinning for Advanced Energy Storage Applications.* 1st ed. Singapore: Springer Singapore; 2021. pp. 201–234.
10. Yeon J, Jang J, Han J, et al. Raman Spectroscopic and X-ray Diffraction Studies of Sulfur Composite Electrodes during Discharge and Charge. *J Electrochem Soc.* 2012;159:1308–1314.
11. Soni CB, Sungjemmenla, Vineeth SK, et al. Patterned Interlayer Enables a Highly Stable and Reversible Sodium Metal Anode for Sodium-Metal Batteries. *Sustain Energy Fuels.* 2023;7:1908–1915.
12. Pu J, Zhong C, Liu J, et al. Advanced: In Situ Technology for Li/Na Metal Anodes: An In-Depth Mechanistic Understanding. *Energy Environ Sci.* 2021;14:3872–3911.
13. Soni CB, Kumar V, Seh ZW. Guiding Uniform Sodium Deposition through Host Modification for Sodium Metal Batteries. *Batter Supercaps.* 2022;5:e264.

14. Luo J, Zhang Y, Matios E, et al. Stabilizing Sodium Metal Anodes with Surfactant-Based Electrolytes and Unraveling the Atomic Structure of Interfaces by Cryo-TEM. *Nano Lett.* 2022;22:1382–1390.
15. Kumar V, Wang Y, Eng AYS, et al. A Biphasic Interphase Design Enabling High Performance in Room Temperature Sodium-Sulfur Batteries. *Cell Reports Phys Sci.* 2020;1:100044.
16. Sungjemmenla, Soni CB, Vineeth SK, et al. Exploration of the Unique Structural Chemistry of Sulfur Cathode for High-Energy Rechargeable Beyond-Li Batteries. *Adv Energy Sustain Res.* 2022;3:2100157.
17. Sungjemmenla, Soni CB, Kumar V. Recent Advances in Cathode Engineering to Enable Reversible Room-Temperature Aluminium–Sulfur Batteries. *Nanoscale Adv.* 2021;3:1569–1581.
18. Sungjemmenla, Vineeth SK, Soni CB, et al. Understanding the Cathode–Electrolyte Interphase in Lithium-Ion Batteries. *Energy Technol.* 2022;10:2200421.
19. Zhao E, Nie K, Yu X, et al. Advanced Characterization Techniques in Promoting Mechanism Understanding for Lithium–Sulfur Batteries. *Adv Funct Mater.* 2018;28:1–21.
20. Geng C, Hua W, Wang D, et al. Demystifying the Catalysis in Lithium–Sulfur Batteries: Characterization Methods and Techniques. *SusMat.* 2021;1:51–65.
21. Rehman S, Pope M, Tao S, et al. Evaluating the Effectiveness of In Situ Characterization Techniques in Overcoming Mechanistic Limitations in Lithium-Sulfur Batteries. *Energy Environ Sci.* 2022;15:1423–1460.
22. Li M, Amirzadeh Z, De Marco R, et al. In Situ Techniques for Developing Robust Li–S Batteries. *Small Methods.* 2018;2:1800133.
23. Zhu W, Liu D, Paolella A, et al. Application of Operando X-ray Diffraction and Raman Spectroscopies in Elucidating the Behavior of Cathode in Lithium-Ion Batteries. *Front Energy Res.* 2018;6:1–16.
24. Chen X, Hou T, Persson KA, et al. Combining Theory and Experiment in Lithium–Sulfur Batteries : Current Progress and Future Perspectives. *Mater Today.* 2019;22:142–158.
25. Xiao F, Wang H, Xu J, et al. Generating Short-Chain Sulfur Suitable for Efficient Sodium–Sulfur Batteries via Atomic Copper Sites on a N,O-Codoped Carbon Composite. *Adv Energy Mater.* 2021;11:1–12.
26. Swinehart DF. The Beer-Lambert law. *J Chem Educ.* 1962;39:333–335.
27. Shiri HM, Aghazadeh M. Synthesis, Characterization and Electrochemical Properties of Capsule-Like NiO Nanoparticles. *J Electrochem. Soc.* 2022;159:E132.
28. Guo Q, Sun S, Kim K, et al. A Novel One-Step Reaction Sodium-Sulfur Battery with High Areal Sulfur Loading on Hierarchical Porous Carbon Fiber. *Carbon Energy.* 2021;3:440–448.
29. Guo Q, Li S, Liu X, et al. Ultrastable Sodium–Sulfur Batteries without Polysulfides Formation Using Slit Ultramicropore Carbon Carrier. *Adv Sci.* 2020;7:1–12.
30. Vineeth SK, Gadhave RV. Corn Starch Blended Polyvinyl Alcohol Adhesive Chemically Modified by Crosslinking and Its Applicability as Polyvinyl Acetate Wood Adhesive. *Polym Bull.* 2023. https://doi.org/10.1007/s00289-023-04746-0.
31. Singh HK, Patil T, Vineeth SK, et al. Isolation of Microcrystalline Cellulose from Corn Stover with Emphasis on Its Constituents: Corn Cover and Corn Cob. *Mater Today Proc.* 2020;27:589–594.
32. Dhawale PV, Vineeth SK, Gadhave RV, et al. Tannin as a Renewable Raw Material for Adhesive Applications: A Review. *Mater Adv.* 2022;3:3365–3388.
33. Vineeth SK, Gadhave RV, Gadekar PT. Investigation of Crosslinking Ability of Sodium Metabisulphite with Polyvinyl Alcohol–Corn Starch Blend and Its Applicability as Wood Adhesive. *Indian Chem Eng.* 2022;64:197–207.

34. Elazari R, Salitra G, Talyosef Y, et al. Morphological and Structural Studies of Composite Sulfur Electrodes upon Cycling by HRTEM, AFM and Raman Spectroscopy Service Morphological and Structural Studies of Composite Sulfur Electrodes upon Cycling by HRTEM, AFM and Raman Spectroscopy. *J Electrochem. Soc.* 2010;157:A1131.
35. Zhu P, Song J, Lv D, et al. Mechanism of Enhanced Carbon Cathode Performance by Nitrogen Doping in Lithium–Sulfur Battery: An X-ray Absorption Spectroscopic Study. *J Phys Chem C.* 2014;118:7765–7771.
36. Li X, Banis M, Lushington A, et al. A High-Energy Sulfur Cathode in Carbonate Electrolyte by Eliminating Polysulfides via Solid-Phase Lithium-Sulfur Transformation. *Nat Commun.* 2018;9:4509.
37. Aquilanti G, Stievano L. X-ray Absorption Near-Edge Structure and Nuclear Magnetic Resonance Study of the Lithium–Sulfur Battery and Its Components. *ChemPhysChem.* 2014;15:894–904.
38. Yan Y, Cheng C, Zhang L, et al. Deciphering the Reaction Mechanism of Lithium–Sulfur Batteries by In Situ/Operando Synchrotron-Based Characterization Techniques. *Adv Energy Mater.* 2019;9:1–14.
39. Liu H, Pei W, Lai WH, et al. Electrocatalyzing S Cathodes via Multisulfiphilic Sites for Superior Room-Temperature Sodium-Sulfur Batteries. *ACS Nano.* 2020;14:7259–7268.
40. Liu D, Zhang C, Zhou G, et al. Catalytic Effects in Lithium – Sulfur Batteries : Promoted Sulfur Transformation and Reduced Shuttle Effect. *Adv Sci.* 2018;5:1700270.
41. Li C, Xi Z, Guo D, et al. Chemical Immobilization Effect on Lithium Polysulfides for Lithium-Sulfur Batteries. *Small.* 2018;14:1701986.
42. Zhang Z, Peng H, Zhao M, et al. Heterogeneous/Homogeneous Mediators for High-Energy-Density Lithium–Sulfur Batteries : Progress and Prospects. *Adv Funct Mater.* 2018;28:1707536:1–23.
43. Chen K, Li HJW, Xu Y, et al. Untying Thioether Bond Structures Enabled by "Voltage-Scissors" for Stable Room Temperature Sodium-Sulfur Batteries. *Nanoscale.* 2019;11:5967–5973.
44. Yui Y, Hayashi M, Nakamura J. In Situ Microscopic Observation of Sodium Deposition/Dissolution on Sodium Electrode. *Sci Rep.* 2016;6:1–8.
45. Seok J, Hyun JH, Jin A, et al. Visualization of Sodium Metal Anodes via Operando X-Ray and Optical Microscopy: Controlling the Morphological Evolution of Sodium Metal Plating. *ACS Appl Mater Interfaces.* 2022;14:10438–10446.
46. Ma B, Lee Y, Bai P. Dynamic Interfacial Stability Confirmed by Microscopic Optical Operando Experiments Enables High-Retention-Rate Anode-Free Na Metal Full Cells. *Adv Sci.* 2021;2005006:1–11.
47. Rychlewski J. On Hooke's law. *J Appl Math Mech.* 1984;48:303–314.
48. Liu X, Wang D, Wan L. Progress of Electrode/Electrolyte Interfacial Investigation of Li-Ion Batteries Via In Situ Scanning Probe Microscopy. *Sci Bull.* 2015;60:839–849.
49. Han M, Zhu C, Ma T, et al. In Situ Atomic Force Microscopy Study of Nano–Micro Sodium Deposition in Ester-Based Electrolytes. *Chem Commun.* 2018;54:2381–2384.
50. Mcmullan D. Scanning electron microscopy 1928–1965. *J Scanning Microscopies.* 1995;17:175–185.
51. Li X, Zhao L, Li P, et al. In-Situ Electron Microscopy Observation of Electrochemical Sodium Plating and Stripping Dynamics on Carbon Nanofiber Current Collectors. *Nano Energy.* 2017;42:122–128.
52. Zeng Z, Barai P, Lee SY, et al. Electrode Roughness Dependent Electrodeposition of Sodium at the Nanoscale. *Nano Energy.* 2020;72:104721.
53. Blanc F, Leskes M, Grey CP. In Situ Solid-State NMR Spectroscopy of Electrochemical Cells: Batteries, Supercapacitors, and Fuel Cells. *Acc Chem Res.* 2013;46:1952–1963.

6 Sodium-sulfur batteries
Similarities and differences with lithium-sulfur battery

C. Sanjaykumar, Rahul Singh, Rajendra Singh, and Vipin Kumar

6.1 NA-S AND LI-S CHEMISTRIES ARE FUNDAMENTALLY DIFFERENT – A MECHANISTIC OVERVIEW

Metal-sulfur batteries have gained academic and industrial attention due to their high theoretical energy density and natural abundance of electrode materials. Considering their higher theoretical gravimetric energy densities (Li-S: 2615 Wh kg^{-1}, Na-S: 1673 Wh kg^{-1}), lithium-sulfur (Li-S) and sodium-sulfur (Na-S) batteries are considered the most extensively explored and viewed as potential contenders [1]. Table 6.1 summarizes the fundamental properties of the electrode materials. Li–S batteries are promising technology for next-generation batteries due to their higher specific capacity (3860 mAh g^{-1}). However, due to the sluggish redox kinetics of sulfur (S_8) and its intermediates, dissolution of polysulfide dendrites, and their energy densities are substantially below their theoretical predictions [2]. Due to the scarcity of Li resources and their high cost, the long-term sustainability of Li-S batteries is questionable. However, sodium-sulfur batteries could be far more affordable compared to Li-S batteries due to the high abundance of sodium in the Earth's crust and water, with relative concentrations of 28,400 mg kg^{-1} and 1000 mgl^{-1} [3]. The prototype of sodium-sulfur was developed way before Li-S batteries. However, due to its high-temperature operation, it could not gain equal academic interest. The feasibility of Na-S chemistry at room temperature was demonstrated in 2007 with polymer electrolyte. The room-temperature operation of these chemistries makes them a promising candidate for long-range electric mobility and large-scale energy storage applications [4].

The cell design of Li-S and room-temperature Na-S batteries are similar, where alkali metal is employed as an anode and a sulfur-carbon composite is used as a cathode, see Figure 6.1a. A polyolefin separator soaked in electrolyte serves as the source of electrolyte and physical barrier. It is worth mentioning that the amount of electrolyte is subjected to the amount of sulfur to ensure the long-term stability of the cell. In addition, for high reversibility, the amount of anode is sensitive toward the loading of the sulfur cathode [5]. For instance, 1 g of sulfur cathode requires about 1.4 g of sodium metal and about 0.43 g of lithium metal to form a balanced RT Na-S

TABLE 6.1
A comparison of the properties of Li and Na metal

S. no.	Key parameters	Lithium metal	Sodium metal	Sulfur
1	Abundance (%)	0.01 or 21 ppm	2.83 or 23,000 ppm	953 ppm
2	Valance	1	1	4
3	Atomic mass (u)	6.29	22.99	32.065
4	Atomic radius (pm)	152	186	180
4	Ionic radius (nm)	0.69	0.98	0.29
5	Density (g cm^{-3})	0.53	0.97	2.07
6	Potential (V vs. SHE)	−3.05	−2.71	NA
7	Ionization energy (kJ/Mol)	520.2	495.8	999.65
8	Gravimetric capacity (mA g^{-1})	3862	1166	1675
9	Volumetric capacity (mA cm^{-3})	2062	1128	NA

and Li-S cell, respectively. However, much more electrolytes and anodes are being employed to fabricate Li-S or RT Na-S cells. Prior attempts indicate that the mass ratio of the anode to cathode (N/P) of about 3–4 is optimal for a stable and reversible operation of the Li-S battery. It could not function stably when employing a similar mass ratio of the anode to the cathode for the RT Na-S battery. Therefore, it is evident that the electrochemical parameters found optimum for Li-S chemistry might not be appropriate for RT Na-S chemistry.

The sulfur cathode is a mixture of sulfur and carbon (sulfur-carbon black or sulfur-Ketjan black or sulfur-mesoporous carbon) in an appropriate weight ratio standard for both Li-S and RT-Na-S batteries. During the discharge process, sulfur undergoes a series of structural changes that involve the formation of soluble polysulfides ($M_2 Sn$ $8 \leq n \leq 3$, M = Li or Na) and insoluble sulfides (MS_2 $8 \leq X \leq 3$, M = Li or Na) in the liquid electrolyte [6]. Despite a similar principle of forming polysulfide, the kinetic and chemical nature of polysulfide varies significantly in Li-S and RT Na-S batteries. For instance, the conversion from Na_2S_8 to Na_2S_6 is relatively sluggish than lithium polysulfides. In addition, two discharge plateaus exist in Li-S batteries corresponding to two different polysulfide states. In contrast, only a single plateau appears in RT Na-S batteries corresponding to lower valent sodium polysulfide. Due to the relatively larger size of sodium ions, the sodium polysulfide is more essential than that of lithium polysulfide, which suggests the cell components must be looked through a prism of the chemical nature of polysulfide [7].

The discharge profile of Li-S and RT Na-S is depicted in Figure 6.1b. Four continuous steps are involved in the compositional and structural conversion of sulfur, sodium polysulfides, and sodium sulfides. The discharging behavior is largely influenced by the physical and chemical properties of Na, which are pretty different from Li metal [8]. The reaction steps involved during discharge are detailed in Table 6.2. Due to sodium polysulfides' high reactivity, their dissolution is way higher than lithium polysulfides in liquid electrolytes. It indicates RT Na-S batteries' separator requirements are more stringent [9].

TABLE 6.2
The electrochemical reaction between sulfur and Na/Li metal anode

Type of battery chemistry	Reaction region	Transition phase
Li-S Batteries	I: $S_8 + 2Li^+ + 2e^- \rightarrow Li_2S_8$	At 2.40 V, a solid-to-liquid phase transition from S_8 to Li_2S_8 occurs in this region, as shown in Figure 6.1. Li_2S_8 dissolved into a liquid electrolyte and started acting as a liquid cathode.
	II: $Li_2S_8 + 2Li^+ + 2e^- \rightarrow 2Li_2S_4$	A liquid-liquid phase reduction occurs in the 2.40 to 2.10 V voltage range, where Li_2S_8 is converted to Li_2S_4 (lower-order polysulfide). When the concentration of polysulfide anion arises and the sulfur-sulfur chain length is shortest, causing a sharp drop in the cell's voltage and a steady increase in the solution's viscosity.
	III: $Li_2S_4 + 2Li^+ + 2e^- \rightarrow 2Li_2S_2$ $Li_2S_4 + 6Li^+ + 6e^- \rightarrow 4Li_2S$	A liquid-solid phase reduction happens in this region, resulting in the dissolution of low-order polysulfide into insoluble Li_2S_2 or Li_2S. The significant capacity of the Li-S cell is derived from this region, which creates the second lower voltage plateau at 2.10 to 1.90 volts.
	IV: $Li_2S_2 + 2Li^+ + 2e^- \rightarrow 2Li_2S$	A solid-solid reduction happening from insoluble Li_2S_2 to Li_2S in this region is between 2.10 and 1.85 V. Due to the non-conductive nature of Li_2S_2 and Li_2S, this process is kinetically sluggish and typically suffers from significant polarization.
Sodium-Sulfur Batteries	I: $S_8 + 2Na^+ + 2e^- \rightarrow Na_2S_8$	This region relates to the voltage plateau at 2.20 V vs. Na/Na$^+$. Cyclo-S_8 changes from solid to liquid in this region and becomes soluble Na_2S_8.
	II: $Na_2S_8 + 2Na^+ + 2e^- \rightarrow 2Na_2S_4$	This region is a sloped voltage zone, with a voltage range of 2.2 and 1.65 V. A liquid-liquid transition from Na_2S_8 to Na_2S_4 occurs in this region.
	III: $Na_2S_4 + \frac{2}{3}Na^+ + \frac{2}{3}e^- \rightarrow 3Na_2S_3$ $Na_2S_4 + 2Na^+ + 2e^- \rightarrow 2Na_2S_2$ $Na_2S_4 + 6Na^+ + 2e^- \rightarrow 4Na_2S$	A liquid transition occurs in this region at the lower plateau at 1.65 V, which is referred to as the short-chain polysulfides (Na_2S_3, Na_2S_2, and Na_2S).
	IV: $Na_2S_2 + 2Na^+ + 2e^- \rightarrow 2Na_2S$	At 1.65–1.2 V, a solid-solid reaction from Na_2S_2 to Na_2S occurs, which leads to the slope section during the discharge process.

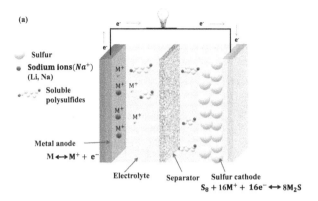
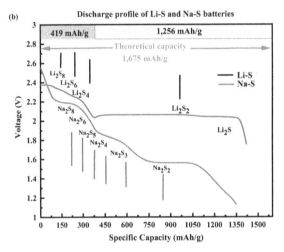
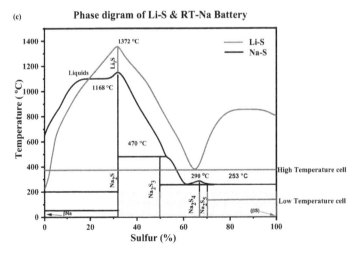

FIGURE 6.1 (a) Schematic diagram of metal sulfur battery. (b) Discharge profiles for Li-S and RT Na-S batteries. (c) Phase diagram of Li-s and RT Na-S batteries.

The binary Li-S and Na-S systems' schematic phase diagrams are shown in Figure 6.1c Li_2S is the only thermodynamically stable binary compound in the Li-S system, whereas Na_2S and several polysulfides (Na_2S_2, Na_2S_4) exist as thermodynamically stable phases in the Na-S system [10]. The Na-S phase diagram also shows the commercial high-temperature cell's operating window and alternative, research-level cell concepts that operate at low temperatures, including room temperature [11].

6.2 POLYSULFIDE SPECIES DISSOLUTION

Polysulfide dissolution is crucial for determining the reversibility and stability of both Li-S and RT-Na-S batteries. In any sulfur-based battery chemistry, polysulfides are the major discharge products. These discharge products reversibly convert to elemental sulfur and polysulfides during cycling. However, after a few tens of cycles, the final desired products, i.e., Na_2S or Li_2S, become irreversible, leading to capacity loss and negatively affecting the cell's stability [12]. In addition, capacity decay is related to the diffusion of sulfur and polysulfides getting dissolved in the electrolyte, resulting in the loss of active material at the anode [13]. The Li-S batteries generally involve a multistep and complicated solid-liquid-solid transition phase mainly accompanied by the shuttle effect. The shuttling phenomena represent sulfur or polysulfides, which are partially or fully reduced from a low charge to a high charge state. These species go back to the cathode and get reduced. During charging, the shuttling phenomena will develop a parasitic current inside the cell. This drastically affects the battery's cycle life and capacity [14]. The solubility of polysulfides is high in RT Na-S battery, allowing the side reaction to happen [15]. To eliminate or suppress the polysulfide shuttling, diffusion, and dissolved polysulfide species in the electrolyte, a combined strategy of tailoring the electrolyte and engineering the sulfur cathode has to be developed [16,17].

6.3 ELECTROLYTE FOR LI-S AND RT NA-S BATTERIES

The electrolyte is the central component of any battery chemistry [18]. Electrolyte salt and solvent are pivotal in Li-S or RT Na-S batteries. For instance, the Li metal anode or Na metal anode is relatively stable with ether-based solvents and $LiPF_6$ or $NaPF_6$ salt. The stability of both the anode is observed to be compromised if the electrolyte salt is changed to $LiClO_4$ or $NaClO_4$. Moreover, the same salt is relatively stable with carbonate-based solvents [19]. Therefore, selecting a suitable salt and solvent is critical and essential to ensure the stability of the metal anode [20]. Besides metal anode, the reactivity of the sulfur cathode with ether or carbonate-based electrolytes poses another challenge. For instance, polysulfide dissolution is way higher in ether-based electrolytes than in carbonate-based electrolytes [21].

From a metal anode perspective, a combination of linear ether (i.e., monoglyme) and cyclic ether (i.e., dioxolane) with LiTFSI or $LiPF_6$ is identified as a suitable electrolyte system for Li-S batteries. However, a similar electrolyte formulation does not promote long-term stability and reversibility to the RT Na-S batteries. For RT Na-S batteries, diglyme-based electrolytes are found to be superior [22].

From a sulfur cathode perspective, an oxygen-containing electrolyte or additive is essential. For example, fluoroethylene carbonate, vinylene carbonate, or dioxolane must be supplied to the electrolyte to ensure the long-term stability of the cell [23]. These electrolyte solvents or additives help bind with polysulfide and mitigate polysulfide cross-over. A combination of either oxolane or fluoroethylene carbonate with ether or carbonate-based solvents is best for Li-S batteries [24]. RT Na-S batteries perform stability with fluoroethylene carbonate in carbonate-based electrolytes; it does not work well with oxolane in ether [25].

The early research on Li-S batteries started with carbonate-based electrolytes; however, the unstable SEI, low Coulombic efficiency, and unwarranted side reactions compelled the researcher to find a better alternative. While the early attempts on RT Na-S batteries were made on gel-polymer or solid polymer electrolytes [26]. Irrespective of the choice of electrolytes, the earlier variant of either Li-S or RT Na-S could barely be cycled for a few tens of cycles. Generally, carbonate electrolytes promote the formation of alkyl-rich SEI, which further reduces to $NaCO_3$ or $LiCO_3$ [27]. To mitigate the number of organic matters, fluorine-containing additives, e.g., FEC (Fluorinated ethylene carbonate), have often been employed [28]. The ether-based electrolytes are more compatible with Na and Li anode than carbonate electrolytes due to improved reduction stability. It promotes relatively higher Coulombic efficiency and minimal stripping/plating overpotential. It facilitates the formation of a thin, compact, and inorganic-rich SEI layer [29]. Though the chemical composition of the SEI significantly depends on the choice of the electrolyte salt, ether-based electrolyte plays a critical role in promoting the formation of a robust SEI layer [30]. As a result, Li-S or RT Na-S batteries exhibit relatively longer cycle life and improved charge transfer kinetics.

6.3.1 Solid Electrolyte Interphase

The SEI layer combines insoluble and partially soluble reduced organic-inorganic species, which develop instantly upon the metal anode's reaction with an organic electrolyte. However, the relative position of energy levels decides the reduction possibilities of the organic electrolyte. The SEI acts as an interface between the anode and the electrolyte. The thickness, uniformity, chemical composition, and mechanical and electrochemical stability of the SEI strongly depend on the nature of the electrolyte and the metal anode. For instance, sodium metal anode reacts vigorously with most organic electrolytes, causing SEI to become thicker and non-uniform. However, a thin, uniform, and compact SEI generally forms on the surface of the lithium metal anode. In addition, the SEI's nature depends on the metal's corrosion rate, kinetic parameters, and half-cell potential of a battery system [31].

Even though the ionic nature of sodium and lithium-ion are similar, the SEI layer for lithium metal anode becomes stable over cycling. At the same time, it continuously grows upon cycling for sodium metal anode [32]. The charge transfer resistance, which is a consequence of the nature of the SEI, is comparatively higher for sodium metal anode. Table 6.3 comprehensively summarizes the chemical constituents of the SEI formed on the surface of lithium and sodium metal.

TABLE 6.3
SEI architecture of lithium and sodium metal anode in a carbonate-based electrolyte

Metal anode	Inorganic species	Organic species
Lithium	Li_2O, Li_2S/Li_2S_2, LiOH, LiF, LiI, Li_3N, Li_2CO_3.	ROLi, RCOOLi, ROCOLi, $RCOO_2Li$, and $ROCO_2Li$ (R = alkyl groups)
Sodium	Na_2O, NaF, and Na_2CO_3	RONa, $ROCO_2Na$, and RCOONa (R = alkyl groups)

TABLE 6.4
Nature of the SEI layer on the lithium and sodium metal anode

	Li metal anode	Sodium metal anode
1	SEI layer – stable after a few cycling	SEI layer- unstable and challenging to predict the timing of formation
2	Solubility of the SEI layer is low	Solubility of the SEI layer is high, leading to instability of SEI
3	Symmetric cell data show lower polarization	Symmetric cell data show higher polarization due to an increase in the interfacial resistance
4	Non-uniform deposition leads to dendrite growth	Non-uniform deposition leads to dendrite growth more toward the flow of electrolytes

The SEI layer of organic species is electrochemically more stable than the inorganic species, and it will restrict and decrease the transport of ions [33]. Compared to Li electrolytes, the solubility of SEI layers is higher for sodium-based electrolytes and augments the instability of the SEI layer on the sodium metal surface [34]. Na metal anode shows inferior electrochemical performance due to a broader distribution of the chemical components of the SEI. In addition, lithium metal anode exhibits lower overpotential/polarization even at higher current density. In contrast, the sodium metal anode shows a higher polarization at even lower currents [30]. The chemical structure of the salt's anion has an indirect relationship with the chemical nature of the SEI. It is essential to understand the formation mechanism of the SEI on the surface of lithium and sodium metal anode [35]. Table 6.4 summarizes the significant differences in the SEI formed on the lithium and sodium metal anodes.

It is known that the polysulfide gets dissolved in the electrolyte and starts to react with the metal anode, leading to a quick capacity fade and cell failure [36]. It is worth highlighting that the inorganic-rich SEI is reported to be more stable under polysulfide attacks than the organic-rich SEI. However, it is immensely difficult to precisely control the chemical composition or the formation process of the SEI [37]. Besides polysulfide poisoning of the anode, the rate of polysulfide release is an additional challenge that requires attention. Due to the larger size of the sodium polysulfides, it is much more critical for RT Na-S batteries [38]. Polysulfide release and poisoning of the anode severely threaten metal-sulfur batteries' stability [39].

6.4 FUTURE PROSPECT

Li-S and RT Na-S cell chemistry is based on the conversion of polysulfides; as a result, both chemistries are sluggish. The insulating nature of the cathode and a high rate of polysulfide dissolution and shuttling cause the cell to fail after a few tens of cycles. The following possibilities could help improve the performance of Li-S and RT Na-S batteries.

A suitable electrolyte system could solve most Li-S and RT Na-S battery issues. The design aspect of polymer electrolytes needs to be revisited to ensure faster charge transfer kinetics. The interfacial resistance needs to be minimized to maintain the cycle life. A suitable polymer host that is unaffected or less affected by the polysulfide attack is imperative. In addition, localized high-concentration electrolytes with higher ionic conductivity and electrochemical stability could be pivotal in mitigating polysulfide dissolution and shuttling. The solid electrolyte that can directly be grown over the surface of the cathode or anode could be groundbreaking for metal-sulfur batteries. The design of the cathode and separator requires modifications to ensure polysulfide entrapping or blocking. To suppress or eradicate the polysulfide dissolution, novel host materials such as porous carbon or inorganic oxides/sulfides/nitrides must be developed. The porous host can be tailored by engineering the physical and chemical structure to ensure polysulfide immobilization. The COFs (Covalent organic framework) could be a promising material system for trapping the polysulfides and providing physical confinement by its inherent nanopores.

Furthermore, internal and external parameters affecting the battery performance must be unveiled, which can be achieved using advanced techniques such as in situ and operando characterizations and theoretical and mathematical models. Screening various anode, cathode, and electrolyte materials using machine learning and artificial intelligence techniques can accelerate the research.

REFERENCES

1. Adelhelm P, Hartmann P, Bender CL, et al. From Lithium to Sodium: Cell Chemistry of Room Temperature Sodium-Air and Sodium-Sulfur Batteries. *Beilstein J Nanotechnol.* 2015;6:1016–1055.
2. Zhao M, Li BQ, Zhang XQ, et al. A Perspective toward Practical Lithium-Sulfur Batteries. *ACS Cent Sci.* 2020;6:1095–1104.
3. Li T, Xu J, Wang C, et al. The Latest Advances in the Critical Factors (Positive Electrode, Electrolytes, Separators) for Sodium-Sulfur Battery. *J Alloys Compd.* 2019;792:797–817.
4. Ye C, Chao D, Shan J, et al. Unveiling the Advances of 2D Materials for Li/Na-S Batteries Experimentally and Theoretically. *Matter.* 2020;2:323–344.
5. Wang L, Wang T, Peng L, et al. The Promises, Challenges and Pathways to Room-Temperature Sodium-Sulfur Batteries. *Natl Sci Rev.* 2022;9:nwab050.
6. Lin L, Zhang C, Huang Y, et al. Challenge and Strategies in Room Temperature Sodium–Sulfur Batteries: A Comparison with Lithium–Sulfur Batteries. *Small.* 2022;18:2107368.

7. Rauh RD, Electrochem J, Rauh RD, et al. A Lithium/Dissolved Sulfur Battery with an Organic Electrolyte. *J Electrochem Soc*. 1979;126:523.
8. Mu P, Dong T, Jiang H, et al. Crucial Challenges and Recent Optimization Progress of Metal–Sulfur Battery Electrolytes. *Energy Fuels*. 2021;35:1966–1988.
9. Abraham KM, Rauh RD, Brummer SB. A Low Temperature NaS Battery Incorporating A Soluble S Cathode. *Electrochim Acta*. 1978;23:501–507.
10. Salama M, Rosy, Attias R, et al. Metal–Sulfur Batteries: Overview and Research Methods. *ACS Energy Lett*. 2019;4:436–446.
11. Sangster J, Pelton AD. The Na-S (Sodium-Sulfur) System. *J Phase Equilibria*. 1997;18:89–96.
12. Sungjemmenla, Soni CB, Vineeth SK, et al. Exploration of the Unique Structural Chemistry of Sulfur Cathode for High-Energy Rechargeable Beyond-Li Batteries. *Adv Energy Sustain Res*. 2022;3:2100157.
13. Bieker G, Küpers V, Kolek M, et al. Intrinsic Differences and Realistic Perspectives of Lithium-Sulfur and Magnesium-Sulfur Batteries. *Commun Mater*. 2021;2:37.
14. Vizintin A, Chabanne L, Tchernychova E, et al. The Mechanism of Li_2S Activation in Lithium-Sulfur Batteries: Can We Avoid the Polysulfide Formation. *J Power Sources*. 2017;344:208–217.
15. Haridas AK, Huang C. Advances in Strategic Inhibition of Polysulfide Shuttle in Room-Temperature Sodium-Sulfur Batteries via Electrode and Interface Engineering. *Batteries*. 2023;9:223.
16. Vineeth SK, Soni CB, Sun Y, et al. Implications of Na-Ion Solvation on Na Anode–Electrolyte Interphase. *Trends Chem*. 2022;4:48–59.
17. Vineeth SK, Tebyetekerwa M, Liu H, et al. Progress in the Development of Solid-State Electrolytes for Reversible Room-Temperature Sodium-Sulfur Batteries. *Mater Adv*. 2022;2: 6415–6440.
18. Xu K. Nonaqueous Liquid Electrolytes for Lithium-Based Rechargeable Batteries. *Chem Rev*. 2004;104:4303–4418.
19. Lee B, Paek E, Mitlin D, et al. Sodium Metal Anodes: Emerging Solutions to Dendrite Growth. *Chem Rev*. 2019;119:5416–5460.
20. Xia L, Yu L, Hu D, et al. Electrolytes for Electrochemical Energy Storage. *Mater Chem Front*. 2017;1:584–618.
21. Liu G, Sun Q, Li Q, et al. Electrolyte Issues in Lithium–Sulfur Batteries: Development, Prospect, and Challenges. *Energy Fuels*. 2021;35:10405–10427.
22. Wang Y, Zhou D, Palomares V, et al. Revitalising Sodium-Sulfur Batteries for Non-High-Temperature Operation: A Crucial Review. *Energy Environ Sci*. 2020;13:3848–3879.
23. Wu J, Liu J, Lu Z, et al. Non-Flammable Electrolyte for Dendrite-Free Sodium-Sulfur Battery. *Energy Storage Mater*. 2019;23:8–16.
24. Aurbach D, Markevich E, Salitra G. High Energy Density Rechargeable Batteries Based on Li Metal Anodes. The Role of Unique Surface Chemistry Developed in Solutions Containing Fluorinated Organic Co-solvents. *J Am Chem Soc*. 2021;143:21161–21176.
25. Zhao X, Zhu Q, Xu S, et al. Fluoroethylene Carbonate as an Additive in a Carbonates-Based Electrolyte for Enhancing the Specific Capacity of Room-Temperature Sodium-Sulfur Cell. *J Electroanal Chem*. 2019;832:392–398.
26. Zhou D, Chen Y, Li B, et al. A Stable Quasi-Solid-State Sodium–Sulfur Battery. *Angew Chemie Int Ed*. 2018;57:10168–10172.
27. Wang H, Wang C, Matios E, et al. Facile Stabilization of the Sodium Metal Anode with Additives: Unexpected Key Role of Sodium Polysulfide and Adverse Effect of Sodium Nitrate. *Angew Chemie*. 2018;130:7860–7863.
28. Kumar H, Detsi E, Abraham DP, et al. Fundamental Mechanisms of Solvent Decomposition Involved in Solid-Electrolyte Interphase Formation in Sodium Ion Batteries. *Chem Mater*. 2016;28:8930–8941.

29. Lee J, Kim J, Kim S, et al. A Review on Recent Approaches for Designing the SEI Layer on Sodium Metal Anodes. *Mater Adv.* 2020;1:3143–3166.
30. Iermakova DI, Dugas R, Palacín MR, et al. On the Comparative Stability of Li and Na Metal Anode Interfaces in Conventional Alkyl Carbonate Electrolytes. *J Electrochem Soc.* 2015;162:A7060–A7066.
31. Peled E. The Electrochemical Behavior of Alkali and Alkaline Earth Metals in Nonaqueous Battery Systems—The Solid Electrolyte Interphase Model. *J Electrochem Soc.* 1979;126:2047–2051.
32. Cheng X-B, Zhang R, Zhao C-Z, et al. A Review of Solid Electrolyte Interphases on Lithium Metal Anode. *Adv Sci.* 2016;3:1500213.
33. Sacci RL, Black JM, Balke N, et al. Nanoscale Imaging of Fundamental Li Battery Chemistry: Solid-Electrolyte Interphase Formation and Preferential Growth of Lithium Metal Nanoclusters. *Nano Lett.* 2015;15:2011–2018.
34. Mogensen R, Brandell D, Younesi R. Solubility of the Solid Electrolyte Interphase (SEI) in Sodium Ion Batteries. *ACS Energy Lett.* 2016;1:1173–1178.
35. Tripathi AM, Su W-N, Hwang BJ. In Situ Analytical Techniques for Battery Interface Analysis. *Chem Soc Rev.* 2018;47:736–851.
36. Dominko R, Vizintin A, Aquilanti G, et al. Polysulfides Formation in Different Electrolytes from the Perspective of X-ray Absorption Spectroscopy. *J Electrochem Soc.* 2018;165:A5014–A5019.
37. Soto FA, Marzouk A, El-Mellouhi F, et al. Understanding Ionic Diffusion through SEI Components for Lithium-Ion and Sodium-Ion Batteries: Insights from First-Principles Calculations. *Chem Mater.* 2018;30:3315–3322.
38. Lee H, Lee JT, Eom K. Improving the Stability of an RT-NaS Battery via In Situ Electrochemical Formation of Protective SEI on a Sulfur-Carbon Composite Cathode. *Adv Sustain Syst.* 2018;2:1800076.
39. Fang R, Xu J, Wang D-W. Covalent Fixing of Sulfur in Metal–Sulfur Batteries. *Energy Environ Sci.* 2020;13:432–471.

7 Other sodium metal-based rechargeable battery technologies
A brief introduction to sodium dual ion batteries

Chhail Bihari Soni, SK Vineeth, and Vipin Kumar

7.1 BIRD-EYE VIEW OF THE SODIUM METAL-BASED BATTERIES

The discovery of sodium-ion conductors, i.e., β-Al_2O_3, lead to the development of high-temperature (HT) sodium-sulfur (Na-S) and sodium metal hydride (Na-MH) batteries [1–3]. These technologies comprise molten sodium metal as an anode, while S/Na_2S_x as the cathode for Na-S [4,5], and metal halide (e.g., $NiCl_2$) cathode with $NaAlCl_4$ as a secondary electrolyte for Na-MH batteries. Due to the poor room-temperature ionic conductivity of β-Al_2O_3 electrolytes, both technologies operate at much higher temperatures (200°C–350°C). The Na-MH battery, also called ZEBRA (zero emission battery research association), is considered a little safer compared to Na-S batteries [6], as the discharge products of the former are comparatively less corrosive than the later technology. The high-temperature operation severely hindered their widespread adoption in stationary storage and mobile applications. Moreover, the construction materials are required to be of high purity and chemical and mechanically strong, leading to a significant increase in the price per kWh.

The sodium-ion battery, which has gained significant attention recently, was attempted in the early 1980s [3]. Interestingly, the concept of sodium-ion and Li-ion batteries emerged around the same point in time. However, due to the better performance of lithium-ion batteries, the interest in developing sodium-ion batteries declined sharply [7]. Sodium-ion batteries regained their position due to a significant jump in the cost of Li-ion battery electrodes and electrolyte materials. With the development of hard-carbon anode, the performance of sodium-ion batteries could be improved remarkably, which led to the development and deployment of sodium-ion batteries.

Though Na-S battery requires about 250°C–350°C to function, the interest remained in developing a cheaper and safer variant of Na-S batteries. A room-temperature variant of Na-S batteries was demonstrated in 2006, where a polymeric membrane was used as an ionic conductor to transport Na^+ ions [8].

DOI: 10.1201/9781003388067-7

Parallelly, other battery technologies, i.e., sodium-air or Na-O_2 comprising sodium metal anode and air or O_2 as cathode, developed recently to increase the energy density [9]. As the superoxides of sodium are relatively stable, the sodium-air batteries were researched hoping to mitigate the instabilities of the products, i.e., superoxides. The present sodium-air batteries can only function with organic electrolytes. More recently, dual-ion batteries have been extended to sodium-carbon systems. Though the battery delivers a little lesser specific capacity than RT Na-S or Na-O_2, it offers higher stability and high operating voltage. The sodium-based dual-ion battery could be the best solution for stationary storage applications.

When combined with the other cathode systems, e.g., O_2, CO_2, or SO_2 and C, the sodium metal batteries produce a high energy density. For instance, the energy density of Na-O_2 batteries is about 1105 Wh/kg, which is at least two times higher than that of the theoretical energy density of Li-ion batteries. The energy density of the Na-CO_2 system is even higher, i.e., 1.13 kWh/kg, which is highly competitive

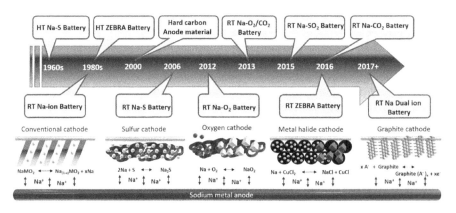

FIGURE 7.1 Timeline of the development of Na metal-based batteries from 1960 to till 2017+ with their working mechanism.

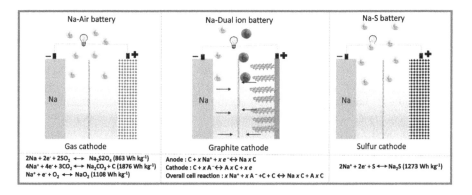

FIGURE 7.2 Schematic illustration of Na-O2 and Na-C battery chemistries. Na-S battery is used for the reference.

with gasoline. The sodium metal coupled with graphite results in a new battery chemistry, i.e., sodium-based dual-ion battery. Compared to the Na-O_2 and Na-CO_2 battery systems, the Na-C battery system is less complex and realizable without significant alteration in the chemistry of the electrode materials. Figure 7.1 shows an overview of various developments in sodium-based battery technologies from 1960 to 2017+ with their working mechanism. Figure 7.2 schematically illustrates the difference between Na-O_2 and Na-C battery chemistries, while Na-S is taken for reference [10–12]. The Na-O_2 and Na-C battery chemistries are detailed in the following sections.

7.2 OVERVIEW OF THE SODIUM-AIR BATTERY

The metal-air batteries are a classic example of a hybrid battery system, comprising a dual design functionality of a conventional battery and an ambient temperature fuel cell [13]. The cathode compartment of the cell serves the function of a fuel cell, while the anode acts like a conventional battery anode. The metal-air promises a huge theoretical energy density, about 10–30 times higher than those of Li-ion chemistries. While there is little control over the choice of the cathode, i.e., air or O_2, the nature of the electrolyte is solely decided by the selection of the anode material. For example, Zn or Al relies on a protic (H^+-based) electrolyte, while Na or Li prefers an aprotic electrolyte.

Sodium-air batteries are considered promising due to their high energy density (~1600 Wh/kg), and a high natural abundance of sodium makes sodium-air batteries more cost-effective (at least 1/3rd) than that lithium-air batteries [14]. Unlike, Na-ion or Na-S battery chemistries, Na-O_2 battery relies on oxygen-peroxide chemistry where the stability of the discharge products decides the reversibility and stability of the cell. Unlike Li-air batteries, the discharge products (i.e., Na_2O) of sodium-air batteries are relatively stable and hence promote electrochemical reversibility during charging reactions. However, sodium-air batteries suffer from high cathodic reaction overpotential during charging, leading to low voltage and energy efficiency. Nonetheless, the sodium-air batteries can be reversibly recharged to 2.9 V vs. Na/Na^+ without apparent side reactions. The major electrochemical reactions of a sodium-air battery are highlighted in Table 7.1.

TABLE 7.1
Basic information of SAB battery, showing the cell reactions, theoretical capacity, and energy density

Cell chemistry	Cell voltage (based on discharge product) (V)	Overpotential (Discharge/charge) (mV)	Theoretical capacity (mAh g^{-1})	The energy density (W h kg^{-1})
$Na^+ + O_2 + e^- \rightarrow$ NaO_2	E^0 ($2NaO_2$) = 2.27 (ΔG^0 = –437.5 kJ mol^{-1})	Discharge < 100 Charging ~ 30–100	1165 (Na) 488 (NaO_2)	1108 (NaO_2) 1605 (Na_2O_2)
$2Na^+ + O_2 + 2e^- \rightarrow$ Na_2O_2	E^0 (Na_2O_2) = 2.33 (ΔG^0 = –449.7 kJ mol^{-1})		689 (Na_2O_2)	

Sodium-air batteries consist of a sodium metal anode and an air-breathing porous cathode separated by a suitable electrolyte. The sodium metal anode undergoes an oxidation reaction during discharging to form Na^+-ions. The Na^+-ion combines with two oxygen ions to create NaO_2. Depending upon the choice of the electrolyte, sodium-air batteries could also be aprotic and protic. In aprotic sodium-air batteries, an organic electrolyte ensures the reversible conduction of Na^+ ions. The discharge products, i.e., NaO_2 and Na_2O_2, which are generally insoluble and non-conductive, get deposited on the porous host, causing kinetics to become sluggish [15]. However, in aqueous sodium-air batteries, a solid oxide Na+-ion conductor, i.e., ($Na_3Zr_2Si_2PO_{12}$), has often been employed [16]. The solid electrolyte ensures high reversibility and stability of the cell [17,18].

The choice of electrolyte is pivotal to ensuring improved reversibility and stability of a sodium-air battery. The electrolyte must remain stable under an oxidative environment; therefore, the electrochemical stability of the electrolytes must be sufficient. Ideally, an electrolyte for SAB should possess a wider electrochemical stability window, low viscosity, chemically stable against superoxide anion (O_2^-), lower vapor pressure, high ionic conduction, and increased oxygen solubility [19]. Furthermore, the dynamics of solvation and desolvation also directly affect the composition of SEI, reversibility, and performance of SABs [18,20]. Solvation dynamics can be analyzed by using ^{23}Na-NMR spectroscopy, a tool for identifying the SEI characteristics of cycled Na metal anode [21]. Commonly, the electrolytes are formed with 1 M of sodium salts, namely $NaCF_3SO_3$ and $NaClO_4$, in organic solvents such as linear ethers (e.g., monoglyme, diglyme, tetra-glyme-ether) and a combination of linear and cyclic carbonates (ethylene carbonate and dimethyl carbonate). Besides that, the studies have predicted that the nature of anion is critical in improving the stability of the solid electrolyte interphase.

From the initial developments of sodium-air batteries, there have been inspiring works from Peled et al. [22]. The battery prototype was first demonstrated in 2011. The cell comprised a liquid Na and E-TEK as anode and air-breathing cathode, respectively, with a polymer electrolyte. To ensure a constant air supply through a porous cathode, the cathode compartment was made with porous membrane. The kinetics of the cell was accelerated by operating it at 105–110°C. In addition, molten anode eradicated the issues such as dendrite formation and sluggish kinetics of air cathode. The cell displayed charging and discharging for 140 cycles with 85% faradaic efficiency. The performance of the cell was also compared with an ionic liquid electrolyte [23,24].

The choice of cathodes also plays a vital role in deciding the performance of a sodium-air battery. During charging and discharging cycles, the kinetics of oxygen-evolution reactions and oxygen-reduction reactions should be quick enough to ensure the reversibility and stability of the cell. Hence, including an electrocatalyst is of utmost importance for enhancing cell performance. The catalyst improves the reversibility of the reactions, as they possess high electrical conductivity and catalyze the complex formation [25,26]. With cycles, insoluble discharge products, such as NaO_2 and Na_2O_2, may accumulate at the cathode surface which obstructs the voids, thereby affecting the diffusion process and, hence, lowering reversibility [27,28]. Therefore the properties of the cathode, such as surface area, porosity, and pore volume, tailor the performance of the sodium-air battery [14]. The general requirement of an air-cathode for metal-air batteries includes a highly interconnected porous structure with diffusion paths for ions and electrons, high surface area, high electrical conductivity, ability to accommodate heteroatom dopants, and chemical and physical stability with cost-effectiveness. Various air cathodes have been reported for SABs; in particular, carbon nanomaterial-based cathodes have been widely reported. CNTs, graphene, carbon nanofibers, and modified carbon nanomaterials received great attention considering their wide attention as they possess high electric conduction and large surface area with porous network structures [29–31].

Binders used in cathode also have an essential role in deciding the performance of sodium-air batteries [7,32,33]. Usually, polymer-based materials are employed as binders with film formation properties, solution processability, and adhesion with various molecules [34–38]. Weak adhesion between a catalyst and cathode or between a cathode and current collector impedes the cycling and reversibility of the cell. Given this, Baek et al. [39] reported catechol-derivative hydrophilic binders for enhancing the interface between the electrocatalyst and the current collector. Better adhesion boosted wettability in the catalyst-current collector interface, preventing delamination of the electrocatalyst. These were confirmed from the morphological analysis and supported by theory calculations. Furthermore, with the augmentation in binding property, cell electrochemical performance improved. Catechol groups in the polymer enhanced binding properties, showing a 15% enhancement in cycling, reduced overpotential, increased oxygen reduction kinetics, and enhanced energy and power efficiency compared to commercial PVDF binder-containing cathode.

In summary, sodium-air battery belongs to a high energy density system, which has enormous application potential. A higher round-trip efficiency and stable charge-discharge plateaus make sodium-air batteries even more attractive over the others. Each battery component is accountable for tuning the performance of the cell. Therefore, a reasonable selection of materials for anode, cathode, electrolyte, and separators is essential. Although the research and development activities on sodium-air batteries are still in their infancy phase, constant efforts are yet to unleash their full potential.

7.3 OVERVIEW OF SODIUM DUAL-ION BATTERIES

Due to the intercalation chemistry of the graphite, it is regarded as a preferred anode for the post-LIBs [40]. Although graphite has the remarkable capacity to host a variety of intercalants, such as Li, K, Cs, and Rb, it is known that the quantity of Na that can be reversibly intercalated into graphite is unexpectedly low (NaC_{186}) [40]. Creating high-performance SDIBs is still a difficult task because there aren't any suitable anode materials for Na insertion. Only a few anode material types have been studied in the SDIBs system. The materials that could store Na in SIBs may theoretically also serve as SDIB anodes. The kinetic matching issue of the anode and cathode must be taken into account since, in contrast to the reaction mechanism of SIBs, Na, and anions react concurrently with the anode and cathode, respectively, in SDIBs. As a result, finding appropriate anode materials is crucial for developing SDIBs. This chapter discusses the usual anodes for SDIBs that have recently been published, including insertion-type materials, alloying-type materials, and conversion-type materials.

7.3.1 OVERVIEW AND HISTORY OF DUAL ION BATTERIES

A battery that can harness the potential of both ions is called a "dual-ion battery." Unlike mono-ion or conventional metal-ion batteries, a dual-ion battery allows reversible intercalation of anions and cations into the cathode and anode during charging reactions, and reverse reactions occur during discharging. Since both anion and cation from the electrolyte partake in the electrochemical reactions, the electrolyte plays a central role in dual-ion batteries. The chemical properties of the electrolyte, i.e., the concentration of ions, change with charge and discharge reactions, which makes a dual-ion battery behave like a Pb-acid battery. The anode could be a layered material that facilitates reversible intercalation/deintercalation of the ions or a metal. The cathode needs to host the anions reversibly, and therefore, it must possess wider interlayer spacing to counter volume change upon cycling. The dual-ion battery generally operates at a higher voltage, allowing it to deliver high energy density, despite the low storage capacity of the electrode materials. Unlike Li-ion or Na-S batteries, the energy and power density of a dual-ion battery can be tuned by altering the concentration of electrolytes.

The first report on the dual-ion system can be traced back to the 1938s when Rüdorff and Hofmann first reported the intercalation of HSO_4^- anions into the graphite [41]. Later in the 1970s, other anions, e.g., ClO_4^- BF_4^-, and $CF_3SO_3^-$, were reported to be reversibly intercalated into the graphite [42]. All the experiments were conducted with an aqueous electrolyte, and therefore, the cell voltage could be increased beyond the decomposition potential of the electrolyte. In 1989, McCullough first reported a formal Li-based dual-ion battery with a nonaqueous electrolyte [43]. In the 2000s, Dahn and co-workers first unveiled the staging process of PF_6^- intercalation into graphite [44]. The formation of a new structure, i.e., $PF_6C_{16,}$ was revealed at sufficiently high voltage, i.e., 5.5 vs. Li/Li$^+$, through in situ XRD. This work paved a path, and the focus has been moved on to dual-ion battery systems. Placke et al. first introduced the term "dual-ion batteries" for reversible intercalation/deintercalation of anions into the graphite, and the same is used today [45].

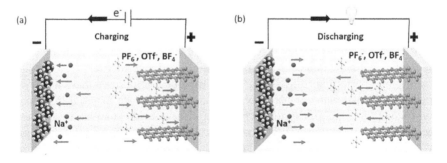

FIGURE 7.3 Schematic illustration of Charging-Discharging process of a sodium dual ion battery.

The device configuration of dual-ion batteries is the same as conventional Li-ion or Na-ion batteries [46]; however, the electrolyte volume is significantly higher than that of traditional cells. The dual-ion batteries differ when it comes to charge-discharge reactions. Unlike the "rocking chair" mechanism, where the cation is the sole performer, the cation and anion reversibly intercalate into the cathode and anode electrode materials. Upon electrochemical reactions, the electrolyte concentration varies in the dual-ion battery; therefore, a relatively higher amount of electrolyte is required. Generally, a thicker separator (at least 10 to 15 times thicker than rocking chair batteries), i.e., 100–150 μm, is used to ensure a higher electrolyte uptake. During charging, Na+-ion intercalates into the anode, and anion (A) intercalates into the cathode, and the reverse occurs during discharge reactions. The corresponding electrochemical process is schematically expressed in Figure 7.3, and reactions are highlighted below [45]:

Anode: $C + xNa^+ + xe^- \leftrightarrow Na_xC$

Cathode: $C + xA^- \leftrightarrow A_xC + xe^-$

Overall cell reaction: $xNa^+ + xA^- + C + C \leftrightarrow Na_xC + A_xC$

Besides Na+-ion, a broader spectrum of other cations, e.g., Li+, K+, Ca^{2+}, Al^{3+}, ionic liquid (IL)-based cations such as 1-butyl-1-methylpyrrolidinium (Pyr14+) or 1-butyl-1 methylpiperidinium (PP14+), follows the similar reaction pathways. However, the anions could be PF_6^-, BF_4^-, imide-based anions (FSI−, FTFSI−, TFSI−, BETI−), ClO_4^-, and DFOB−, etc.

7.3.2 Developments of sodium dual ion batteries

Sodium-based dual-ion batteries are a promising and sustainable alternative to the Li-based dual-ion batteries. Sodium-based dual-ion batteries are in their infancy, and constant efforts are in progress to develop state-of-the-art anode and cathode materials. Besides that, attempts are made to design suitable electrolytes for sodium-based dual-ion batteries. The advancement in developing cathode, anode, and electrolyte systems is detailed in the following sections.

7.3.2.1 Modifications in cathode materials

A relatively high voltage in dual-ion batteries can be ascribed to the intercalation/deintercalation of anions into/from cathode materials during the charging/discharging reactions. Anion intercalation and deintercalation are significantly influenced by the architectures and properties of cathode materials [47]. Due to its distinct layered structure, environmental friendliness, and affordability, graphite is often employed as the cathode in sodium-based dual-ion batteries [48]. The electrochemistry of anion intercalation depends critically on the structure, morphologies, and size of graphite [49]. The degree of graphitization and crystallinity can both be used to describe the graphite's structure. Note that the intercalation of anions into the graphite is more sensitive toward the degree of graphitization and crystal perfection of graphite than the intercalation of Na+-ions. An ordered, free-from defects and imperfections structure of graphite often leads to store more amount of the anions, which is contrary to that of the storage of Na+-ions [50].

Organic materials with aromatic components have garnered massive attention as the prospective cathode material for sodium-based dual-ion batteries. The high durability, diverse functionality, and tuneable electrical properties make them a superior candidate to accommodate anions [42].

7.3.2.2 Anode materials

The host materials that could reversibly store/intercalate the Na$^+$-ions can be employed as anode materials in sodium-based dual-ion batteries. The sodium metal anode, which is generally utilized as an anode in metal batteries [51–54], can also be used as an anode for sodium-based dual-ion batteries. The plating/stripping behaviors of sodium metal during the charging/discharging process, i.e., xNa$^+$ + xe$^-$ ↔ Na, where Na is the metallic sodium, and x is their corresponding valence, provide the basis of their storing mechanism. It ensures that when employing the sodium metal as the anode, a polymer or quasi-polymer is generally used as an electrolyte to ensure improved safety [55].

Carbon-based materials, for instance, hard carbon, can also be employed as an anode. For example, pine needles (PNC) derived hard carbon is demonstrated as a potential anode for sodium-based dual-ion batteries [56]. The storage mechanism in hard carbon differs from the sodium metal, as the former stores the charges through a "house-of-cards" model. However, the nature of sodium storage is mainly depending on the potential. For instance, the sodium storage below 0.5 V vs. Na/Na$^+$ favors stripping/plating instead of intercalation behavior. Besides that, soft carbon, which could withstand structural changes, has also been examined for its sodium storage performance [57].

In addition to carbonaceous materials, other insertion-type materials have also been explored as anodes in sodium-dual ion batteries. For instance, the open structure of TiO$_2$ makes it a suitable intercalation-type anode for sodium-based dual-ion batteries. Attempts are made to fabricate a dual-ion cell with TiO$_2$ anode and graphite cathode. However, due to the disparity in the intercalation kinetics of the Na+-ion and PF6- anion into TiO$_2$ and graphite, the battery system could not remain stable for longer cycles. The intercalation kinetics of Na$^+$-ions is way slower in TiO$_2$ structure [58]. Therefore, improving the Na diffusion kinetics by altering the morphology or structure of TiO$_2$ may further improve the electrochemical performance of TiO$_2$-graphite dual-ion batteries.

The alloying type anode materials have also been tested for their ability to store sodium reversibly. The energy storage of alloying-type materials is based on electrochemical alloying/dealloying reactions. For example, tin (Sn) is employed for its ability to form an alloy with sodium. Sn undergoes alloying (Na$_x$Sn$_y$) and dealloying (Sn) reactions upon electrochemical reactions. More recently, a sodium-based dual-ion battery has been demonstrated using an Sn-based anode and graphite-based cathode [59]. The following electrochemical reactions were proposed to occur on the anode and cathode:

Anode: Sn + Na$^+$ + e$^-$ ↔ NaSn
Cathode: xC + PF$_6^-$ ↔ C$_x$(PF$_6$) + e$^-$
Overall cell reaction: Sn + xC + Na$^+$ + PF$_6^-$ ↔ NaSn + C$_x$(PF$_6$)

TABLE 7.2
A recent summary of the electrochemical performance of SDIBs

Electrode material	Electrolyte	Working mechanism	Rate capability	Cyclic performance
P@C//graphite [60]	1 M NaPF6 in EC/DMC/EMC, 10% FEC	Alloying reaction	120.6 at 1000 mA g^{-1}	201.5 at 0.25 A g^{-1} after 140 cycles
Sn//EG [61]	1 M NaPF6 in EC/DMC/EMC	Alloying reaction	91.6 at 1200 mA g^{-1}	94 at 500 mA g^{-1} after 600 cycles (cathode)
Sn//graphite [59]	1 M NaPF6 in EC/DMC/EMC, 10% FEC	Alloying reaction	61 at 500 mA g^{-1}	70 at 200 mA g^{-1} after 400 cycles (cathode)
FePO$_4$//graphite [62]	1 M NaPF$_6$ in PC/EMC	Insertion		111.8 at 0.2 A g^{-1} after 250 cycles
Na$_2$Ti$_3$O$_7$@G//coronene [63]	1 M NaPF$_6$ in EC/DEC	Insertion	60 at 1000 mA g^{-1}	80 at 500 mA g^{-1} after 5000 cycles
TiO$_2$//graphite [58]	1 M NaPF$_6$ in EC/EMC	Insertion	102 at 1500 mA g^{-1}	98 at 500 mA g^{-1} after 1400 cycles
Carbon molecular sieve/KS6 graphite [64]	1 M NaPF$_6$ in EC/EMC	Insertion	110 at 2000 mA g^{-1}	150 at 500 mA g^{-1} after 500 cycles
Hard carbon//graphite [65]	2.55 M NaTFSI in TMP	Insertion		34.5 at 500 mA g^{-1} after 200 cycles (cathode)
Phosphorus-doped soft Carbon//graphite [66]	1 M NaPF$_6$ in EC/DMC	Insertion	73 at 3000 mA g^{-1}	81 at 1000 mA g^{-1} after 900 cycles
Soft carbon/graphite [57]	1 M NaPF$_6$ in EC/DMC	Insertion	40 at 2000 mA g^{-1}	54 at 1000 mA g^{-1} after 800 cycles
Hard carbon//graphite [67]	0.8 M NaPF$_6$ in PC	Insertion	46 at 558 mA g^{-1}	53 at 186 mA g^{-1} after 200 cycles (cathode)
Hard carbon//KS6 graphite [56]	1 M NaPF$_6$ in EC/EMC	Insertion	98 at 1000 mA g^{-1}	127 at 500 mA g^{-1} after 1000 cycles
SnP$_2$O$_7$//KS$_6$ graphite [68]	1 M NaPF$_6$ in EC/DMC/EMC	Conversion	65 at 3000 mA g^{-1}	70 at 2 A g^{-1} after 1000 cycles
MoS$_2$@C//graphite [69]	1 M NaPF$_6$ in EC/DMC	Conversion	63.6 at 2000 mA g^{-1}	90.5 at 500 mA g^{-1} after 500 cycles
MoS$_2$@C//EG [70]	1 M NaPF$_6$ in EC/DMC/EMC	Conversion	45 at 2000 mA g^{-1}	40 at 1 A g^{-1} after 500 cycles (cathode)
MoS$_2$@C//EG [71]	1 M NaPF$_6$ in EC/DMC/EMC	Conversion	38.5 at 1000 mA g^{-1}	60 at 100 mA g^{-1} after 300 cycles (cathode)
MoS$_2$@C//graphite [72]	1 M NaPF$_6$ in EC/DMC/EMC	Conversion	35 at 1000 mA g^{-1}	55 at 200 mA g^{-1} after 200 cycles (cathode)
WS$_2$/C@CNTs//EG [73]	1 M NaPF$_6$ in EC/DMC/EMC	Conversion	160 at 1000 mA g^{-1}	150 at 1000 mA g^{-1} after 500 cycles

During charging, while PF_6^- intercalates into the graphite layers to generate graphite intercalation compounds (GICs), Na^+-ions undergo alloying reactions with Sn to form Na-Sn alloy. Though reactions are reversible, the volume expansion causes the anode to pulverize upon repeated cycling. Table 7.2 comprehensively summarizes the performance of the anode materials used for sodium-based dual-ion batteries.

7.4 FUTURE PROSPECTS

Dual-ion storage allows the cell to operate at a high voltage. However, it causes most electrolytes to decompose. It must be noted that the electrolyte acts as the active material for a dual-ion battery, and its weight must be considered while determining the overall specific capacity or specific energy density. The capacity of the cell can easily be altered by changing the concentration of the electrolyte. Despite that, it is yet to be identified as an optimal recipe for the electrolyte. A graphite cathode with increased interplanar spacing must be designed to enable a high-energy cell. Hence, the active cathode material (particle size) and electrode structure (appropriate porosity) must be tuned for anion uptake/release. In addition, to extend the life of the cell, a thorough understanding of the interphase and anion-cation crosstalk is a must.

Only a few materials have been studied as anodes for sodium-based dual-ion batteries. Therefore, it is essential to identify many more anode materials with exceptional electrochemical performance. Understanding the kinetics of the anodic reaction is crucial to enhance the performance further. The creation of new anode materials, optimization of the electrochemical performance, and prediction of these materials could all be benefited from the integration of theoretical calculations and experimental characterizations. Furthermore, it is necessary to monitor the real-time changes that occur in the electrode materials' structure. In situ characterization techniques, for instance, in situ XRD, in situ TEM, in situ Raman spectroscopy, or a combination of techniques, must be developed to gain insight into the mechanistic aspects of the anion storage.

REFERENCES

1. Palomares V, Serras P, Villaluenga I, et al. Na-Ion Batteries, Recent Advances and Present Challenges to Become Low-Cost Energy Storage Systems. *Energy Environ Sci.* 2012;5:5884–5901.
2. Palomares V, Casas-Cabanas M, Castillo-Martínez E, et al. Update on Na-Based Battery Materials. A Growing Research Path. *Energy Environ Sci.* 2013;6:2312–2337.
3. Delmas C. Sodium and Sodium-Ion Batteries: 50 Years of Research. *Adv Energy Mater.* 2018;8:1703137.
4. Hueso KB, Armand M, Rojo T. High Temperature Sodium Batteries: Status, Challenges and Future Trends. *Energy Environ Sci.* 2013;6:734.
5. Soni CB, Sungjemmenla, Vineeth SK, et al. Challenges in Regulating Interfacial-Chemistry of the Sodium-Metal Anode for Room-Temperature Sodium-Sulfur Batteries. *Energy Storage.* 2022;4:e264.

6. Kim B-R, Jeong G, Kim A, et al. High Performance Na-CuCl$_2$ Rechargeable Battery toward Room Temperature ZEBRA-Type Battery. *Adv Energy Mater*. 2016;6:1600862.
7. Sungjemmenla, Vineeth SK, Soni CB, et al. Understanding the Cathode–Electrolyte Interphase in Lithium-Ion Batteries. *Energy Technol*. 2022;10:2200421.
8. Park CW, Ahn JH, Ryu HS, et al. Room-Temperature Solid-State Sodiumsulfur Battery. *Electrochem Solid-State Lett*. 2006;9:123–126.
9. Hartmann P, Bender CL, Vračar M, et al. A Rechargeable Room-Temperature Sodium Superoxide (NaO$_2$) Battery. *Nat Mater*. 2013;12:228–232.
10. Sungjemmenla, Soni CB, Kumar V. Recent Advances in Cathode Engineering to Enable Reversible Room-Temperature Aluminium–Sulfur Batteries. *Nanoscale Adv*. 2021;3:1569–1581.
11. Sungjemmenla, Soni CB, Vineeth SK, et al. Exploration of the Unique Structural Chemistry of Sulfur Cathode for High-Energy Rechargeable Beyond-Li Batteries. *Adv Energy Sustain Res*. 2022;3:2100157.
12. Sungjemmenla, Soni CB, Vineeth SK, et al. Unveiling the Physiochemical Aspects of the Matrix in Improving Sulfur-Loading for Room-Temperature Sodium–Sulfur Batteries. *Mater Adv*. 2021;2:4165–4189.
13. Liu S, Liu S, Luo J. Carbon-Based Cathodes for Sodium-Air Batteries. *New Carbon Mater*. 2016;31:264–270.
14. Chawla N, Safa M. Sodium Batteries: A Review on Sodium-Sulfur and Sodium-Air Batteries. *Electronics*. 2019;8:1201.
15. Yadegari H, Sun X. Recent Advances on Sodium–Oxygen Batteries: A Chemical Perspective. *Acc Chem Res*. 2018;51:1532–1540.
16. Das SK, Lau S, Archer LA. Sodium-Oxygen Batteries: A New Class of Metal-Air Batteries. *J Mater Chem A*. 2014;2:12623–12629.
17. Vineeth SK, Tebyetekerwa M, Liu H, et al. Progress in the Development of Solid-State Electrolytes for Reversible Room-Temperature Sodium-Sulfur Batteries. *Mater Adv*. 2022;2:6415–6440.
18. Vineeth SK, Soni CB, Sun Y, et al. Implications of Na-Ion Solvation on Na Anode–Electrolyte Interphase. *Trends Chem*. 2022;4:48–59.
19. Yadegari H, Sun X. Sodium–Oxygen Batteries: Recent Developments and Remaining Challenges. *Trends Chem*. 2020;2:241–253.
20. Raguette L, Jorn R. Ion Solvation and Dynamics at Solid Electrolyte Interphases: A Long Way from Bulk? *J Phys Chem C*. 2018;122:3219–3232.
21. Lutz L, Alves Dalla Corte D, Tang M, et al. Role of Electrolyte Anions in the Na–O$_2$ Battery: Implications for NaO$_2$ Solvation and the Stability of the Sodium Solid Electrolyte Interphase in Glyme Ethers. *Chem Mater*. 2017;29:6066–6075.
22. Peled E, Golodnitsky D, Mazor H, et al. Parameter Analysis of a Practical Lithium- and Sodium-Air Electric Vehicle Battery. *J Power Sources*. 2011;196:6835–6840.
23. Peled E, Golodnitsky D, Hadar R, et al. Challenges and Obstacles in the Development of Sodium-Air Batteries. *J Power Sources*. 2013;244:771–776.
24. Sun Q, Yang Y, Fu Z-W. Electrochemical Properties of Room Temperature Sodium–Air Batteries with Non-Aqueous Electrolyte. *Electrochem commun*. 2012;16:22–25.
25. Yin W, Fu Z. The Potential of Na–Air Batteries. *ChemCatChem*. 2017;9:1545–1553.
26. Cheng F, Chen J. Metal–Air Batteries: From Oxygen Reduction Electrochemistry to Cathode Catalysts. *Chem Soc Rev*. 2012;41:2172.
27. Ha S, Kim J-K, Choi A, et al. Sodium-Metal Halide and Sodium-Air Batteries. *ChemPhysChem*. 2014;15:1971–1982.
28. Lin X, Sun Q, Doyle Davis K, et al. The Application of Carbon Materials in Nonaqueous Na-O$_2$ Batteries. *Carbon Energy*. 2019;1:141–164.

29. El-Kady MF, Shao Y, Kaner RB. Graphene for Batteries, Supercapacitors and Beyond. *Nat Rev Mater.* 2016;1:16033.
30. Nomura A, Ito K, Kubo Y. CNT Sheet Air Electrode for the Development of Ultra-High Cell Capacity in Lithium-Air Batteries. *Sci Rep.* 2017;7:45596.
31. Pendashteh A, Palma J, Anderson M, et al. Doping of Self-Standing CNT Fibers: Promising Flexible Air-Cathodes for High-Energy-Density Structural Zn–Air Batteries. *ACS Appl Energy Mater.* 2018;1:2434–2439.
32. Gadhave RV, Vineeth SK, Mahanwar PA, et al. Effect of Addition of Boric Acid on Thermo-Mechanical Properties of Microcrystalline Cellulose/Polyvinyl Alcohol Blend and Applicability as Wood Adhesive. *J Adhes Sci Technol.* 2021;35:1072–1086.
33. Eng AYS, Nguyen D-T, Kumar V, et al. Tailoring Binder–Cathode Interactions for Long-Life Room-Temperature Sodium–Sulfur Batteries. *J Mater Chem A.* 2020;8:22983–22997.
34. Vineeth SK, Gadhave RV, Gadekar PT. Polyvinyl Alcohol–Cellulose Blend Wood Adhesive Modified by Citric Acid and Its Effect on Physical, Thermal, Mechanical and Performance Properties. *Polym Bull.* 2022. https://doi.org/10.1007/s00289-022-04439-0.
35. Dhawale PV, Vineeth SK, Gadhave RV, et al. Tannin as a Renewable Raw Material for Adhesive Applications: A Review. *Mater Adv.* 2022;3:3365–3388.
36. Gadhave RV, Vineeth SK. Synthesis and Characterization of Starch Stabilized Polyvinyl Acetate-Acrylic Acid Copolymer-Based Wood Adhesive. *Polym Bull.* 2022. https://doi.org/10.1007/s00289-022-04558-8.
37. Vineeth SK, Gadhave RV. Corn Starch Blended Polyvinyl Alcohol Adhesive Chemically Modified by Crosslinking and Its Applicability as Polyvinyl Acetate Wood Adhesive. *Polym Bull.* 2023. https://doi.org/10.1007/s00289-023-04746-0.
38. Vineeth SK, Gadhave RV, Gadekar PT. Investigation of Crosslinking Ability of Sodium Metabisulphite with Polyvinyl Alcohol–Corn Starch Blend and Its Applicability as Wood Adhesive. *Indian Chem Eng.* 2022;64:197–207.
39. Baek MJ, Choi J, Wi TU, et al. Strong Interfacial Energetics between Catalysts and Current Collectors in Aqueous Sodium-Air Batteries. *J Mater Chem A.* 2022;10:4601–4610.
40. Xu J, Dou Y, Wei Z, et al. Recent Progress in Graphite Intercalation Compounds for Rechargeable Metal (Li, Na, K, Al)-Ion Batteries. *Adv Sci.* 2017;4:1700146.
41. Rüdorff W, Hofmann U. Über Graphitsalze. *Z Anorg Allg Chem.* 1938;238:1–50.
42. Sui Y, Liu C, Masse RC, et al. Dual-Ion Batteries: The Emerging Alternative Rechargeable Batteries. *Energy Storage Mater.* 2020;25:1–32.
43. McCullough FP, Levine CA, Snelgrove RV. *United States Patent 4,830,938.* 1989:US4830938A.
44. Seel JA, Dahn JR. Electrochemical Intercalation of PF[sub 6] into Graphite. *J Electrochem Soc.* 2000;147:892.
45. Placke T, Fromm O, Lux SF, et al. Reversible Intercalation of Bis(trifluoromethanesulfonyl)imide Anions from an Ionic Liquid Electrolyte into Graphite for High Performance Dual-Ion Cells. *J Electrochem Soc.* 2012;159:A1755–A1765.
46. Sundaram PM, Soni CB, Sungjemmenla, et al. Reviving Bipolar Construction to Design and Develop High-Energy Sodium-Ion Batteries. *J Energy Storage.* 2023;63:107139.
47. Li W-H, Ning Q-L, Xi X-T, et al. Highly Improved Cycling Stability of Anion De-/Intercalation in the Graphite Cathode for Dual-Ion Batteries. *Adv Mater.* 2019;31:1804766.
48. Wu M, Liao J, Yu L, et al. 2020 Roadmap on Carbon Materials for Energy Storage and Conversion. *Chemistry.* 2020;15:995–1013.
49. Kravchyk KV, Kovalenko MV. Rechargeable Dual-Ion Batteries with Graphite as a Cathode: Key Challenges and Opportunities. *Adv Energy Mater.* 2019;9:1901749.

50. Wang J, Tu J, Lei H, et al. The Effect of Graphitization Degree of Carbonaceous Material on the Electrochemical Performance for Aluminum-Ion Batteries. *RSC Adv.* 2019;9:38990–38997.
51. Soni CB, Kumar V, Seh ZW. Guiding Uniform Sodium Deposition through Host Modification for Sodium Metal Batteries. *Batter Supercaps.* 2022;5:e202100207:1–8.
52. Eng AYS, Soni CB, Lum Y, et al. Theory-Guided Experimental Design in Battery Materials Research. *Sci Adv.* 2022;8. https://doi/10.1126/sciadv.abm2422.
53. Soni CB, Sungjemmenla, Vineeth SK, et al. Patterned Interlayer Enables a Highly Stable and Reversible Sodium Metal Anode for Sodium-Metal Batteries. *Sustain Energy Fuels.* 2023;7:1908–1915.
54. Soni CB, Arya N, Sungjemmenla, et al. Microarchitectures of Carbon Nanotubes for Reversible Na Plating/Stripping Toward the Development of Room-Temperature Na–S Batteries. *Energy Technol.* 2022;10:2200742.
55. Xu X, Lin K, Zhou D, et al. Quasi-Solid-State Dual-Ion Sodium Metal Batteries for Low-Cost Energy Storage. *Chem.* 2020;6:902–918.
56. Wang X, Zheng C, Qi L, et al. Carbon Derived from Pine Needles as a Na^+-Storage Electrode Material in Dual-Ion Batteries. *Glob Challenges.* 2017;1:1700055.
57. Fan L, Liu Q, Chen S, et al. Soft Carbon as Anode for High-Performance Sodium-Based Dual Ion Full Battery. *Adv Energy Mater.* 2017;7:1–8.
58. Wang X, Qi L, Wang H. Anatase TiO_2 as a Na^+-Storage Anode Active Material for Dual-Ion Batteries. *ACS Appl Mater Interfaces.* 2019;11:30453–30459.
59. Sheng M, Zhang F, Ji B, et al. A Novel Tin-Graphite Dual-Ion Battery Based on Sodium-Ion Electrolyte with High Energy Density. *Adv Energy Mater.* 2017;7:1601963.
60. Wang H. High-Performance Phosphorus – Graphite Dual-Ion Battery. *ACS Appl Mater Interfaces.* 2019;11:45755–45762.
61. Xie D, Zhang M, Wu Y, et al. A Flexible Dual-Ion Battery Based on Sodium-Ion Quasi-Solid-State Electrolyte with Long Cycling Life. *Adv Funct Mater.* 2020;30:1–7.
62. Li C, Wang X, Li J, et al. $FePO_4$ as an Anode Material to Obtain High-Performance Sodium-Based Dual-Ion Batteries. *Chem Commun.* 2018;54:4349–4352.
63. Dong S, Li Z, Rodríguez-Pérez IA, et al. A Novel Coronene//$Na_2Ti_3O_7$ Dual-Ion Battery. *Nano Energy.* 2017;40:233–239.
64. Wang X, Qi L, Wang H. Commercial Carbon Molecular Sieves as a Na^+-Storage Anode Material in Dual-Ion Batteries. *J Electrochem Soc.* 2017;164:A3649–A3656.
65. Jiang X, Liu X, Zeng Z, et al. A Nonflammable Na^+-Based Dual-Carbon Battery with Low-Cost, High Voltage, and Long Cycle Life. *Adv Energy Mater.* 2018;8:1–9.
66. Ma R, Fan L, Chen S, et al. Offset Initial Sodium Loss to Improve Coulombic Efficiency and Stability of Sodium Dual-Ion Batteries. *ACS Appl Mater Interfaces.* 2018;10:15751–15759.
67. Hu Z, Liu Q, Zhang K, et al. All Carbon Dual Ion Batteries. *ACS Appl Mater Interfaces.* 2018;10:35978–35983.
68. Mu S, Liu Q, Kidkhunthod P, et al. Molecular Grafting towards High-Fraction Active Nanodots Implanted in N-Doped Carbon for Sodium Dual-Ion Batteries. *Natl Sci Rev.* 2021;8:nwaa178.
69. Cui C, Wei Z, Xu J, et al. Three-Dimensional Carbon Frameworks Enabling MoS_2 as Anode for Dual Ion Batteries with Superior Sodium Storage Properties. *Energy Storage Mater.* 2018;15:22–30.
70. Li Z, Yang L, Xu G, et al. Hierarchical MoS_2@N-Doped Carbon Hollow Spheres with Enhanced Performance in Sodium Dual-Ion Batteries. *ChemElectroChem.* 2019;6:661–667.

71. Liu Y, Hu X, Zhong G, et al. Layer-By-Layer Stacked Nanohybrids of N,S-Co-Doped Carbon Film Modified Atomic MoS_2 Nanosheets for Advanced Sodium Dual-Ion Batteries. *J Mater Chem A*. 2019;7:24271–24280.
72. Zhu H, Zhang F, Li J, et al. Penne-Like MoS_2/Carbon Nanocomposite as Anode for Sodium-Ion-Based Dual-Ion Battery. *Small*. 2018;14:1703951.
73. Liu YJ, Li J, Liu B, et al. Confined WS_2 Nanosheets Tubular Nanohybrid as High-Kinetic and Durable Anode for Sodium-Based Dual Ion Batteries. *ChemSusChem*. 2022;16:e202201200.

8 Conclusion and prospects

Vipin Kumar

8.1 CONCLUSION AND PROSPECTS

The book systematically reviewed the history and progress made in developing Na-S battery chemistries at high temperatures (200°C–300°C), intermediate temperatures (100°C–200°C), and room temperature (30°C). It provides the reader with excellent state-of-the-art knowledge of room-temperature sodium-sulfur batteries. Room-temperature (RT Na-S) batteries are still in their infancy stage, and the performance of a practical battery is not compared with other commercial systems. RT Na-S batteries offer the opportunity to develop a high-energy, long-life, and cost-effective long-term (6–8 hours) energy storage solution.

Chapter 2 reviews the history and status of the sodium metal anode. It provides comprehensive information about the approaches and strategies to modify or construct a stable sodium metal anode for RT Na-S batteries. The bottleneck challenges in realizing a stable and reversible sodium metal anode are also highlighted. This topic is actively considered as one of the critical problems. In particular, forming a stable SEI is the key to addressing numerous problems simultaneously. A chemically and mechanically stable SEI is effective in suppressing dendrite growth. A thin, compact, and uniform SEI is desirable to extend the stability of the sodium metal anode. Volume expansion upon stripping/plating of sodium causes loss of sodium and thereby decay in the Coulombic efficiency. The amount of sodium is critical in deciding the stability of the RT Na-S batteries. Theoretically, a sodium loading of about 1.4 g is sufficient to function RT Na-S battery; however, in practice, much higher sodium is required to realize the long-term stability of the cell. On the one hand, a higher sodium loading favors the long-term operation of the cell. On the other hand, it raises safety concerns.

Chapter 3 provides the reader with an excellent survey and review of the state of knowledge of sulfur cathodes. A high theoretical capacity and relatively positive redox potential of sulfur made it an attractive cathode candidate. Irrespective of the operating temperature, the sulfur cathode undergoes polysulfide reactions. On the one hand, polysulfide is undesirable for the safe operation of the cell. On the other hand, it is required to realize a higher capacity of the cell. For instance, the higher-order polysulfide (Na_2S_8–Na_2S_6) does not limit the cycle life of the cell; however, it contributes a little to the capacity build-up. Though conversion of the long-chain polysulfide to short-chain polysulfide helps improve the capacity of the cell, it deteriorates the cycle life of the cell. The conductive carbon hosts that accommodate elemental sulfur and polysulfide could somewhat boost cell performance. It is not yet sufficient to consider for practical applications. This research topic is under active investigation, and the emphasis is given to designing a novel sulfur cathode with higher sulfur loading without causing polysulfide dissolution and shuttling.

The electrolyte alone could solve many problems of RT Na-S batteries. Unlike ceramic electrolytes, non-aqueous liquid electrolyte exhibits higher room-temperature ionic conductivity, excellent electrode wetting, faster charge transfer kinetics, and ease of preparation. Despite all these merits, liquid electrolytes are accountable for polysulfide dissolution and shuttling phenomenon. This book comprehensively discussed the importance of electrolytes for RT Na-S batteries. Besides polysulfide dissolution and cross-over, the solvation dynamics play a vital role in determining the stability of the cell. For instance, a liquid electrolyte generally consists of free-ion and contact-ion pairs, and these ion pairs' relative concentration strongly affects electrolyte properties.

Moreover, the nature of ion pairs decides the fate of the SEI, i.e., free-ion pairs favor the formation of an organic-rich SEI. In contrast, an electrolyte with contact-ion pairs of aggregate-ion pairs forms an inorganic-rich SEI. The high-concentration or localized high-concentration electrolytes are the recent examples of contact-ion pair or aggregate-ion pairs electrolytes.

The performance of RT Na-S battery critically depends on the physical and chemical properties of the sodium, sulfur, and electrolyte and their mutual interaction. In the ex situ techniques, where the test environment is different from the cell environment, a convincing model to govern the dendrite growth, SEI growth, polysulfide formation, and dissolution could not be realized. Decent progress in studying SEI and dendrite growth has been made in the past few years; there is a shortage of literature on the formation and dissolution of sodium polysulfide. In situ techniques offer a broader spectrum to analyze electrode materials' chemical and physical properties. The interaction of the materials and the source is a matter of concern.

The cell chemistry of RT Na-S shares similarities with Li-S batteries; however, due to the high basicity of the sodium polysulfides, the polysulfide attack in RT Na-S batteries is slightly more severe than in Li-S batteries. The local chemistry of the electrolyte for Li-S and RT Na-S is more or less similar. Yet, the charge/discharge kinetics of RT Na-S chemistry is much slower, leading to higher polarization losses. Therefore, the knowledge gathered for Li-S batteries may or may not be applied to study RT Na-S batteries.

Sodium metal batteries, for instance, sodium-air and dual-ion batteries, are the recent addition under the umbrella of metal anode batteries. The sodium-air battery exhibits a high energy density, which is way higher than Li-ion or Na-ion batteries; the reversibility and long-term stability are questionable. However, sodium-based dual-ion is in its infancy stage, and more intensive research needs to be conducted to compare its performance with the existing sodium chemistries or Li-ion chemistries.

The strategies for developing high-energy RT Na-S batteries must consider the depreciation of the electrodes' weight to prioritize safety. The weight ratio of electrolyte to sulfur loading or sulfur to sodium loading must be reconsidered carefully to extend the cycle life further. The progress presented in this book is encouraging to researchers, academicians, and students. RT Na-S batteries' future looks promising as the next generation high-energy rechargeable battery beyond Li chemistries.

Index

ab initio molecular dynamics (AIMD) 66, 86, 88, 96
acceptor number (AN) 79
activated carbon nanofiber (AC-CNF) 51
activated ultra microporous coffee carbon (ACC) 45
active material 1–2, 9–10, 42, 45–46, 48, 52, 56, 59, 66, 88–89, 98, 131, 146
additives 11, 24–25, 85, 88, 92, 102–104, 132
adhesion 54, 98, 103, 141
adsorption energy (E_{ad}) 44, 48, 50, 67
advanced characterization techniques 12, 65
alkali metal anode 23, 28, 79, 123
alucone 29
aluminium foil 33
aluminium-based MOF (MIL-53(Al)) 100
ambient temperature 5, 7, 18–19, 39, 58, 96, 98, 104, 139
amorphous 46, 50, 99–100
angstrom 123
anion 24, 26, 79, 86, 89–90, 92, 104, 129, 140, 142–144, 146
annealing 60, 63, 96, 98
anode 24–35, 39–40, 42–43, 79, 81, 85–86, 88–92, 98, 101–104, 116, 118, 121–123, 127–129, 131–134, 137–144, 146, 151
anode to cathode ratio (N/P) 128
aprotic 139–140
aqueous media 85
architecture 5–6, 12, 42, 48, 51, 53, 59, 67, 133, 143
Arrhenius model 99
artificial intelligence 105, 134
artificial interface 23
atomic force microscopy (AFM) 23, 123
atomic force microscopy-based scanning electrochemical microscopy (AFM–SECM) 66
atomic layer deposition (ALD) 28, 62
atomic radii 21, 96, 128
atomic weight 40, 48, 128
auxiliary 122
average mass density 56

ball milling 58, 96
barium titanium trioxide ($BaTiO_3$) 50
battery performance 9–10, 21, 24, 84, 115, 134
beryllium tetrafluoride ion 142–143
beta alumina solid electrolyte (BASE) 86
binder 1, 9, 24, 44, 56, 58, 64, 141

binding energy 50, 67, 97, 121
biphasic interface (dual components) 3
bipyramid prism 51
bis(2,2,2-trifluoromethyl) ether (BTFE) 27, 92
bis(fluor sulfonyl)imide (FSI) 24, 143
bis-N, N'-propyl-4,9-dicarboxamidediamantane (DCAD) 103
bis(trifluoro methane)sulfonimide (TFSI) 24, 26
blade casting 58
blending 93, 99–100
body centred cubic structure (BCC) 21
boiling point 40, 81
boron 47, 101
bottle necks 41, 79, 85, 89, 151
brine 17
brittle 30
1-bromopropane 25
buffer layer 21

calcium cation (Ca^{2+}) 143
capacity 85, 88, 92–93, 98–101, 103–104, 117, 138–139, 141–147, 151
capacity decay 12, 59, 86, 88, 89, 92–93, 102, 131
capacity fade 2, 41, 45, 66, 102, 133
capillary adsorption 57
carbon 6, 9, 11, 28, 32–33, 44–49, 52–53, 60–64, 66–67, 98, 121, 127–128, 134, 137–138, 141, 144–145, 151
carbon black 9, 128
carbon dioxide (CO_2) 138–139
carbon disulfide (CS_2) 52
carbon felt 33, 44
carbon fiber network (CFC) 54, 60
carbon fibre cloth (CFC) 34, 43, 52, 60
carbon fluoride (CF) 121
carbon matrix 39, 43, 46, 50–52, 63, 66
carbon molecular sieve 145
carbon monoxide (CO) 21, 86
carbon nanobubble 46
carbon nanofibre 46, 61, 141
carbon nanosphere 48, 61
carbon nanotube (CNT) 49, 62
carbon paper 29
carbon scaffold 45, 48, 52
carbon surface 44
carbonate electrolyte 20, 24, 103, 132
carbonization 49, 61–63
carboxyl groups 9
carboxymethyl Cellulose (CMC) 9
catalysis 47–48, 50–51
catalytic effect 47, 49–51, 67

153

cathode electrolyte interface (CEI) 11, 102, 121
catholyte 2, 51–52, 98
cell failure 41, 79, 133
cell performance 10, 79, 85, 102, 105, 120, 124, 140, 152
cell resistance 20
ceramic solid electrolyte 2
cetyltrimethylammonium ammonium bromide (CTAB) 26
charge density 44
charge transfer kinetics 2, 9, 132, 134, 152
chemical compatibility 79, 89
chemical composition 22, 65, 103, 116, 119–121, 132–133
chemical oxidation 62
chemical vapour deposition 60
chemisorbed 46
chloride-rich electrolyte 121
cobalt (Co) 2, 48–49, 61, 120
cobalt (CoS_2) 51, 63
coin cell 10, 21
cold pressed 59
computational model 65–67
conductive filler 1, 39
conductive host 43, 57–58
confinement 45–46, 52, 121, 134
contact ion pair (AGG) 89–90, 92
continental crust 2
contraction 21
conventional wet method 58
conversion reaction 32, 39, 49, 66, 97
co-ordination sites 99
copolymerization 58, 99
copper (Cu) 17, 48, 89
coronene 145
corrosion rate 132
corrosive 5, 83, 103, 137
cost effective 1, 39, 50, 139, 151
coulombic efficiency (CE) 19–21, 23, 27, 32, 43, 45, 53, 85, 87–88, 95, 99, 102, 132, 151
counter electrode 91, 123
covalent fixing 52
covalent organic framework (COF) 134
cracks 9, 20–21, 59
critical evaluation 105
crosslinking 53, 93
cryo-transmission electron microscopy (Cryo-TEM) 116
crystal ion pairs (CIP) 89–90
crystalline 31, 99–100
crystallographic changes 65
current collector 5, 9, 17, 23, 28, 33, 58–59, 83, 89, 92, 141
current density 20, 26–27, 29–34, 46, 50, 60–64, 97–98, 102, 133
cycling life 12, 43

3D printing 59
defects 18, 28, 44, 53, 93, 96, 143
degradation 18, 42, 53, 55, 86, 115, 123
degree of graphitization 143
delamination of electrocatalyst 141
dendrites 10, 12, 18, 20–21, 25–29, 32–33, 102, 116–117, 122–123, 127
density functional theory (DFT) 51, 66–67, 85
depth profiling 116
design 28, 41, 44, 48–50, 55–56, 59, 66–67, 89, 94, 121–123, 127, 134, 139, 143
deterioration 42, 49
determinant factor 10
dielectric constant 10, 25, 80–81, 85, 92
dielectric permittivity 85
diethyl carbonate (DEC) 26–27, 30–33, 60, 62–63, 81, 86, 95, 145
diethylene glycol dimethyl ether (Diglyme) 24, 26, 29, 32, 34, 60, 80, 87, 89, 103–104, 121, 131, 140
differential scanning calorimetry (DSC) 100–101
diffraction peak 46, 119
diffusion barrier 25, 28
diffusion coefficient 53
1,2-dimethoxyethane (DME) 23, 26–27, 89–92, 104
dimethyl acetamide (DMA) 80
dimethyl carbonate (DMC) 81, 84
1,3-dioxolane (DOL) 23
direct-hopping mechanism 96
discharge potential 5, 53
discharge products 4–7, 9, 42, 45, 80, 89, 103, 119, 137, 139–140
discharge profile 7, 28, 128, 130
dissolution 2, 9–11, 27, 31–34, 41–45, 48, 51, 53–55, 57, 65–67, 86, 88–89, 97, 99, 101–102, 104, 115–117, 119, 122, 124, 127–129, 131, 134, 152
donor number (DN) 79–80, 104
double-shell carbon structure 46
drip method 58
dual-ion batteries 138, 141–144, 146, 152

electric field 10, 20, 23, 80, 116
electric insulator 39
electric vehicles 2
electroactivity 43
electrochemical cell 122–123
electrochemical energy storage 1
electrochemical performance 9, 26, 29, 33, 47, 52–53, 55, 59, 65, 67, 84, 89, 93, 97–98, 102, 133, 141, 144–146
electrochemical stability window (ESW) 10, 80–81, 88, 100, 102
electrode electrolyte interface (EEI) 11, 85–86, 102

Index

electrolysis 17
electrolyte decomposition 30, 104, 115
electrolyte/sulfur (E/S) ratio 10, 55–56, 67
electron beam 116, 123
electron migration 59
electron transfer 45
electronegativity 40, 47, 80, 96
electronic conductivity 9, 43, 50
electrospinning 60, 62, 64, 93
elemental doping 11
encapsulation 45, 48, 50, 58
energy density 1, 3–6, 9–10, 12, 18, 39, 42, 52, 55–56, 80–81, 86, 103, 127, 138–139, 141–142, 146, 152
energy efficiency 4, 139
energy storage device 79
etching time 45
ether based electrolyte 88
ethyl methyl carbonate (EMC) 19, 81, 95, 145
ethylaluminum dichloride 103
ethylene carbonate (EC) 81, 84, 140
1-ethyl-3-methylimidazolium bis(fluorosulfonyl) imide (EMMIFSI) 103
evaporation 9, 17
electron microscopy 116, 123
ex-situ techniques 117–118
extrinsic interphase 25, 28
extrusion type 59

face centred cubic structure (FCC) 21
Fermi energy 102
fibrous structure 46
field emission scanning electron microscopy (FESEM) 49, 51, 118, 123
flame retardants 11, 24, 88
flammability 11, 88
flash point 10–11, 25, 81
flat geometry 89
fluoroethylene carbonate (FEC) 24, 27, 132
fluoride-rich electrolyte 121
flux 10, 18, 20, 29, 62
Fourier-transform infrared (FTIR) spectroscopy 86, 89, 91, 93, 97, 118
free-ion pairs 152
fuel cell 139
functional filler 11, 101–102
functional group 52, 88, 99, 119–120

gas evolution 11, 21, 86, 115
gasoline 139
gel-polymer electrolyte (GPE) 88, 93–94, 97
Gibbs free energy 21
glass transition temperature (T_g) 100
glassy electrolyte 99
gold (Au) 48, 62
grafting 93
grain boundary 96

graphene 30, 54, 63–64, 100, 141
graphene aerogel 48, 61
graphene nanosheets 47, 61
graphene oxide 28, 54
graphite 139, 141–146
gravimetric energy density 4
ground-breaking 51, 134

hard carbon 32, 144–145
hardness 28, 34
heat treatment 58, 61–62, 64
heteroatom 46–47, 141
high concentrated electrolytes (HCEs) 25, 88–92, 94
high dielectric constant 10, 80
high energy applications 12
high energy density 52, 55, 80–81, 138–139, 141–142, 152
high polarity 10
high specific energy density 39
high surface area 9, 33, 45, 59, 141
high temperature sodium sulfur battery (HT Na-S) 2–6, 12, 81, 83, 86, 99
high voltage 40, 58, 142–143, 146, 152
highest occupied molecular orbital (HOMO) 80–82, 86, 92
high-order polysulfide 42, 119, 151
hollow carbon nanobubbles on porous carbon nanofibers (CHNBs@PCNFs) 44
hollow matrices 51
hollow nanocages 49
hollow nanosphere 46, 60
homogeneous distribution 29, 51
Hooks law 123
host matrix 57, 92, 99–100
hot pressed 59
house-of-card 144
hybrid approach 124
hydration process 85
hydro fluoro ethers 90, 92
hydrocarbon 21
hydrochloric acid (HCL) 45, 62
hydrogen bond 85
hydrogen fluoride (HF) 23, 88
hydrogen sulfate (HSO_4^-) 142
hydrothermal 62
hydroxide 19

impedance 33, 43, 92
indium iodide (InI_3) 24–25
inelastic scattering 119, 120
inorganic components 22, 24, 92
inorganic solid-state electrolytes (ISEs) 96, 98–99
in-situ analysis 12
in-situ pyrolysis 63
insulating nature 11, 41, 43, 116, 120, 134

intercalation 1, 141–144, 146
interfacial contact 52, 98
interfacial engineering 10–11
interlaced mesoporous carbon hollow
    nanospheres (iMCHS) 44, 46
intermediate temperature 5, 151
internal stress 21
intrinsic interphase 25
investigation 66, 92, 122, 152
ion selective membrane 11
ion transfer 27, 48
ionic conductivity 5–6, 10–11, 28, 30, 80–81,
    84–85, 88, 91, 93–94, 96, 98–101,
    134, 152
ionic liquid (IL) 29, 98, 103, 140, 143
ionic mobility 10–11, 85, 88, 92, 96, 99–100
ionic nature 132
ionic radius 128
ionization energy 128
iron oxide (Fe_2O_3) 34
iron(III) phosphate ($FePO_4$) 145
isothermal 60

Ketjan black 9, 128
kinetic Monte Carlo (KMC) 66
kinetics parameter 132
knock-off mechanism 96
KS6 graphite 145

lead (II) fluoride (PbF_2) 96
Lewis's acidity/basicity 10, 80
liquid nitrogen 116
lithiated graphite (LiC_6) 79
lithium bis(trifluoromethanesulfonyl)imide
    (LiTFSI) 23, 131
lithium carbonate ($LiCO_3$) 132
lithium carbonate (Li_2CO_3) 133
lithium fluoride (LiF) 133
lithium hexafluorophosphate ($LiPF_6$) , 131
lithium hydroxide (LiOH) 121, 133
lithium iodide (LiI) 133
lithium ion (Li+) 133
lithium ion batteries (LIBs) 1, 2, 86, 88, 92,
    100, 141
lithium metal 127–128, 92, 132–133
lithium oxide (Li_2O) 133
lithium perchlorate ($LiClO_4$) 131
lithium sulfide (Li_2S) 66, 129, 131, 133
lithium-sulfur (Li-S) 18, 53, 55–56, 65,
    127–134, 152
localized high concentrated electrolyte (LHCE)
    25, 88, 90, 92, 94
long chain polysulfides , 9, 41–42, 47, 50, 52, 86,
    119–120, 151
lowest unoccupied molecular orbital (LUMO) 23,
    80–81, 86, 102, 104
low-order polysulfide 42, 129

machine learning (ML) 105, 134
magnesium (Mg) 18, 30
magnesium oxide (MgO) 99
magnetic stirrer 57
manganese oxide (MnO_2) 54
mechanical failure 21
mechanism 12, 41, 47–49, 65–66, 80, 96, 99, 102,
    116–117, 119, 138–139, 141, 143–145
melting point 3, 40, 81
metal deposition 116
metal halide 3, 137
metal nanocluster 48
metal oxides 39, 50, 58
metal sulfides 50–51
metal-alloy interface (MAI) 31, 103
metallic organic framework (MOF) 49, 51, 61–62,
    95, 100
metal-sulfur batteries 1, 39, 40–41, 45, 50, 55,
    79–80, 127, 134
microporous carbon 9, 45, 60, 104, 119
microscopic tool 116
microsphere 46, 48, 60, 62
mobility of ions 80, 88
modules 3
moisture 23, 96, 117
molecular dynamics (MD) 85–86
molecular energy level 79
molecular interactions 11, 67
molecular layer deposition (MLD) 28–29
molecular simulation 105
molybdenum sulfide (MoS_x) 50
monochromatic light 119–120
monomer 53, 97
morphological 42, 122–123, 141
multi-walled carbon nanotubes (MWCNT) 51
MXene host 67

nanodots 62
nano/micro filler-reinforced matrices 100
National Aeronautics and Space Administration
    (NASA) 3, 5, 84
NAtrium SuperIonic CONductors (NASICON) 5,
    32, 81, 95, 98
N-butyl-N- methylpyrrolidinium
    bis(fluorosulfonyl)imide (Pyr14FSI) 98
NGK Insulator 2–3, 81
nickel (Ni) 48–49, 61
nickel (II) chloride ($NiCl_2$) 4, 137
nickel foam 33
niobium disulfide (NbS_2) 67
nitrogen 47–51, 61, 63, 101, 116
nitrogen-doped nickel hollow spheres CNFs
    (Ni-NCFs) 47–48
N-Methylpyrrolidone (NMP) 9
nonaqueous 2, 59, 142, 152
noncompatible 89
nonconductive 50, 123, 129, 140

Index

non-destructive 118
nonflammability 103
nonmetal dopants 50, 47
nonpolar 46, 67
nonporous 47
nonsolvent 59
nontoxicity 41
non-uniformity 10, 20, 25, 85, 116, 132–133
nuclear magnetic resonance (NMR) spectroscopy 104, 118, 140
nucleation 18, 20–21, 33–34, 118, 122
nucleophilic 79, 102

oligoether units 100
open ring sizes (ORS) 67
optical microscopy 116, 118, 122
organic rich phase 121
organic solvents 6, 21, 58, 79, 102, 140
organo sulfide 53
overpotential 20–21, 27, 29, 31–34, 92, 98, 102–104, 122, 132–133, 139, 141
oxidation durability 90
oxidation potential 23–24, 66, 103
oxidation state 39–40, 54, 120–121
oxolane 132
oxygen-evolution reactions (OER) 140
oxygen-peroxide 139
oxygen-reduction reactions (ORR) 140
oxynitrides 24

passivation layer 19, 27, 30, 104
Pb-acid batteries 3, 4, 142
pentaerythritol tetraacrylate (PETEA)–tris [2- (acryloyloxy)ethyl] isocyanurate (THEICTA) 93, 95
perchlorate (PF_6) 24, 142–144, 146
Per-dew Burke Ernzerhof (PBE) 67
perfluoro-2-methyl-3-pentanone Petrochemical by-product 2
phase diagram 130–131
phase inversion 59, 93
phosphidation 62
phosphorous 47, 52, 98, 145
phosphorus pentasulfide (P_2S_5) 24, 98, 103
photosynthesis 1
physical confinement 11, 46, 52, 134
physical mixing 57, 60, 63
pine needles (PNC) 144
plasma-enhanced atomic layer deposition (PE-ALD) 28
plasticizer 92–93, 99, 102
polar groups 9
poly methyl methacrylate (PMMA) 100
poly vinylidene fluoride (PVDF) 9, 95, 100, 141
poly(butyl methacrylate) (PBMA) 100
poly(vinylidene fluoride-co-hexafluoropropylene) PVDF-HFP 93, 95, 100

polyacrylonitrile (PAN) 53–54, 64, 101
polyaniline (PANI) 54, 64
polyatomic structures 39
polydopamine (PDA) 64
polyethylene glycol (PEG) 95
polyethylene glycol dimethyl ether (PEGDME) 60
polyethylene oxide (PEO) 2, 84, 87, 95, 100–101
polyhedral oligomeric silsesquioxanes (POSS) 95
polyhedron 45
polyolefin separator 9, 127
polypyrrole (PPy) 54, 64
polysulfides (Na_2S_8 to Na_2S) 2, 5–6, 9, 40–55, 64–67, 86, 88–89, 93, 97–98, 101, 104, 117–119, 121, 128–131, 134, 152
polyvinyl alcohol (PVA) 100
polyvinylpyrrolidone (PVP) 100
poor affinity 57
poor reactivity 120
porosity 28, 58–59, 140, 146
porous carbon microsphere (PCM) 46
porous geometry 93
potassium bis(trifluoromethyesulfonyle) imide (KTFSI) 24, 26
potassium ion (K^+) 143
potential-dependent rates 66
pouch cell 21, 28
pre-processing 58
projector augmented wave (PAW) 67
propane (C_3H_8) 21, 86
propylene carbonate (PC) 19, 21, 27, 29–30, 32–34, 60–64, 81, 85–87, 91, 93, 95, 97, 145

quantum chemistry 86
quasi solid state electrolyte 93
quenching 96

radial species 86
Raman microscopy 66
Raman spectroscopy 115, 118–120, 146
rate capability 48, 145
rational design 50
reaction time 116
recrystallization 57
redox potential 1, 17, 28, 151
redox reaction 66
reduced graphene oxide (rGO) 28, 54
reduction process 23
reflux method 62
renewable energy 1
reversibility 5, 26, 30, 46, 53, 85, 100, 119, 127, 131, 139–141, 152
rhombic dodecahedron 45
rock mining 17
rocking chair mechanism 143

room temperature sodium sulfur battery (RT Na-S) 2–3, 5, 127–128, 130–132, 134, 138, 151–152
round trip efficiency 141

sands model 20
scanning electron microscopy (SEM) 44, 89, 91, 103, 118, 123
selected Area Electron Diffraction (SAED) patterns 44, 46
selenium (Se) 80, 98
selenium sulfide 53
selenization 63
self-decomposition 18
self-discharge 42–43
self-weaving 51
semi-quantitative analysis 120
short chain polysulfides 41, 45, 48, 52, 117, 119, 129, 151
short-range interaction 80
shuttling 2, 9–11, 41–42, 45–46, 50, 66, 85–86, 88, 98, 101–102, 124, 131, 134, 152
side reactions 10, 22–23, 25, 29, 45, 89, 99, 115, 132, 139
silicon dioxide (SiO_2) 29, 66, 95, 99
silver (II) sulfide (Ag_2S) 93
single metal atom 47–48
sintering 96
sluggish kinetics 2, 11, 41, 46, 140
slurry 9, 58–59
soaking 93
soda ash (Na_2CO_3) 17
sodiation 41, 54
sodium Alginate ($NaC_6H_7O_6$) 9
sodium benzenedithiolate (PhS_2Na_2)- rich layer 29
sodium bis(fluorosulfonyl)imide (NaFSI) 26–27, 64, 82, 87, 89–92, 94–95, 100–101, 121
sodium bis(trifluoromethylsulfonyl)imide $NaN(SO_2CF_3)_2$
sodium bromide (NaBr) 25, 30–31
sodium chloride (NaCl) 22, 103
sodium deposition 18, 20, 85, 103, 116, 122–123
sodium fluoride (NaF) 22–23, 133
sodium hexafluorophosphate ($NaPF_6$)
sodium hydroxide (NaOH) 122
sodium ion (Na+) solvation 84
sodium metal 18–25, 27–29, 31, 33–35, 39, 80, 82–83, 85, 88–90, 94, 98, 102, 116, 121, 123, 128, 132–133, 137, 139–140, 144, 151–152
sodium oxide (Na_2O) 22–23, 121–122, 133, 139
sodium perchlorate ($NaClO_4$) 27, 29–30, 34, 60–64, 82, 84, 87, 95, 98, 104, 121, 131, 140
sodium peroxide (Na_2O_2) 139–140

sodium polysulfide (Na_2S_6) 8, 41, 52, 54, 97, 118–120, 128–129, 151
sodium sulfide (Na_2S) 5, 7–8, 24, 41–42, 46–47, 50–51, 61, 86, 93, 98, 118–120, 129, 131
sodium thio-orthophosphate (Na_3PS_4) 96
Sodium trifluoromethanesulfonate (NaOTf) 26, 80, 82, 103–104, 121
sodium-air battery ((Na-O_2) 138–141, 152
sodium-metal hydride (Na-MH) 137
soft carbon 144–145
sol-gel method 98, 61–62
solid electrolyte interface (SEI) 85–86, 88, 92, 102–104, 115–116, 121–124, 132–134, 140, 151
solid metal batteries (SMBs) 80–81, 85, 89, 92–93, 98, 103
solid polymer electrolytes (SPE) 99–102
solid-state electrolytes (SSEs) 99
solvation dynamic 11, 66, 85, 104, 140, 152
solvation number 85
solvent ink 59
solvent molecules 10–11, 79–81, 83, 85–86, 89, 92, 104
solvent-free electrostatic spray deposition (ESD) 58
solvent-separated ion pair (SSIP) 89–90
specific discharge capacity 1, 99
spongy texture 52
stainless steel 89, 104
stannous chloride ($SnCl_2$) 24–25, 27, 64, 103
stress 18, 21, 23, 28, 98
stripping/plating 21, 132, 144, 151
structural integrity 42, 46
sulfur dioxide (SO_2) 138
sulfur loading 9–10, 45–46, 50, 55–56, 59, 152
sulfurized polyacrylonitrile (SPAN) 53, 63–64, 88
superoxide anion (O_2^-) 140
surface area 9, 33, 45, 59, 140–141
surface chemistry 117, 121
surface energy 18
sustainable 1, 143
symmetric cell 88, 92–93, 98, 103, 123, 133

tellurium (Te) 53, 64, 67
tetra ethylene glycol dimethyl ether (TEGDME) 26, 60–62, 64, 84, 99
tetrafluoroethy tetrafluoro propyl ether (TTE) 24, 92
tetrahydrofuran (THF) 89
thermal drying 58
thermal insulator 39
thermal runway 12, 18, 83
thermal stability 25–26, 43, 99
thermodynamic integrity 18
thermoplastic polyurethane (TPU) 95
thioether 52–53, 63, 121

Index

tin (II) pyrophosphate (SnP_2O_7) 145
titanium dioxide (TiO_2) 50, 100
Tokyo Electric Power Company (TEPCO) 2
toxic solvent 57
transference number 6, 10, 25
transition metal oxides 39
transition region 6–7, 40–41
transmission electron microscopy (TEM) 115, 117, 123
transmission X-ray microscopy (TXM) 65
trifluoromethanesulfonate (OTf) 24
trimethyl phosphate 88
trisodium sulfanylidene(trisulfido)-lambda5-phosphane (Na_3PS_4) 29, 96, 98
tubular cell 3
Tungsten (W) 63
Tungsten disulfide (WS_2) 67, 145

ultra sonic device 57
ultraviolet–visible spectroscopy 115
uniform deposition 120
uniform distribution 45, 53

vacuum 57, 58, 64, 116, 123
valance 128
Van der Waals 46, 52, 59
vanadium sulfide (VS_2) 67
vibrational modes 120
Vienna ab initio simulation package (VASP) 67
vinylene carbonate (VC) 104
viscosity 10–11, 25, 27, 55, 57, 80–81, 85, 89–92, 94, 129, 140

Vogel-Tammann-Fulcher (VTF) model 99
voids 46, 57, 140
volatile 11, 58, 89
voltage hysteresis 49
voltage plateaus 40–41
voltage scissors 53
volume expansion 9–11, 21, 42, 146, 151
volumetric energy density 56
volumetric fluctuations 42, 44, 46, 53

wavelength of light 119
weak adsorption 43
wet-solvothermal strategy 52
wettability 10, 11, 21, 25, 27, 28, 88–90, 92, 141
wrapping 58

X-ray absorption near edge structure (XANES) 117–118, 120
X-ray absorption spectroscopy (XAS) 65–66
X-ray diffraction (XRD) 44, 65–66, 101, 116–119, 142, 146
X-ray fluorescence microscopy (XRF) 65
X-ray microscopy (XRM) 65
X-ray photoelectron spectroscopy (XPS) 116
X-ray tomographic microscopy (XTM) 65

zeolitic imidazolate framework (ZIF) 61
zero emission battery research association (ZEBRA) 3, 17, 137
zinc (Zn) 32
zinc sulfide (ZnS) 51, 62, 120
zirconium dioxide (ZrO_2) 99

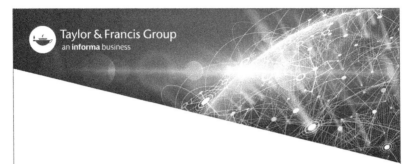